Library of
Exact Philosophy

LEP

Editor:
Mario Bunge, Montreal

Co-editors:
Sir Alfred Jules Ayer, Oxford
Rudolf Carnap †, Los Angeles, Calif.
Herbert Feigl, Minneapolis, Minn.
Victor Kraft, Wien
Sir Karl Popper, Penn

Springer-Verlag New York Wien

Library of Exact Philosophy 9

Charles Castonguay

Meaning and Existence in Mathematics

Springer-Verlag New York Wien 1972

Printing type: Sabon Roman
Composed and printed by Herbert Hiessberger, Pottenstein
Binding work: Karl Scheibe, Wien
Design: Hans Joachim Böning, Wien

ISBN 0-387-81110-9   Springer-Verlag New York - Wien
ISBN 3-211-81110-9   Springer-Verlag Wien - New York

General Preface to the LEP

The aim of the Library of Exact Philosophy is to keep alive the spirit, if not the letter, of the Vienna Circle. It will consequently adopt high standards of rigor: it will demand the clear statement of problems, their careful handling with the relevant logical or mathematical tools, and a critical analysis of the assumptions and results of every piece of philosophical research.

Like the Vienna Circle, the Library of Exact Philosophy sees in mathematics and science the wellsprings of contemporary intellectual culture as well as sources of inspiration for some of the problems and methods of philosophy. The library of Exact Philosophy will also stress the desirability of regarding philosophical research as a cooperative enterprise carried out with exact tools and with the purpose of extending, deepening, and systematizing our knowledge about human knowledge.

But, unlike the Vienna Circle, the Library of Exact Philosophy will not adopt a school attitude. It will encourage constructive work done across school frontiers and it will attempt to minimize sterile quarrels. And it will not restrict the kinds of philosophical problem: the Library of Exact Philosophy will welcome not only logic, semantics and epistemology, but also metaphysics, value theory and ethics as long as they are conceived in a clear and cogent way, and are in agreement with contemporary science.

Montreal, January 1970

Mario Bunge

Preface

The take-over of the philosophy of mathematics by mathematical logic is not complete. The central problems examined in this book lie in the fringe area between the two, and by their very nature will no doubt continue to fall partly within the philosophical remainder. In seeking to treat these problems with a properly sober mixture of rhyme and reason, I have tried to keep philosophical jargon to a minimum and to avoid excessive mathematical complication. The reader with a philosophical background should be familiar with the formal syntactico-semantical explications of proof and truth, especially if he wishes to linger on Chapter 1, after which it is easier philosophical sailing; while the mathematician need only know that to "explicate" a concept consists in clarifying a heretofore vague notion by proposing a clearer (sometimes formal) definition or formulation for it. More seriously, the interested mathematician will find occasional recourse to EDWARD's *Encyclopedia of Philosophy* (cf. bibliography) highly rewarding. Sections 2.5 and 2.7 are of interest mainly to philosophers.

The bibliography only contains works referred to in the text. References are made by giving the author's surname followed by the year of publication, the latter enclosed in parentheses. When the author referred to is obvious from the context, the surname is dropped, and even the year of publication or "ibid." may be dropped when the same publication is referred to exclusively over the course of several paragraphs. In some quotations I have converted the notation of the original to suit my own.

This book developed from a thesis presented to the Faculty of Graduate Studies of McGill University in 1971. Both thesis and book owe much to the encouragement of MARIO BUNGE, who deftly and patiently stimulated my growing interest in these philosophical problems. The content and form of the book have also benefited from suggestions of PHILIP OLIN and JOHN TRENTMAN, and, especially, WALTER BURGESS and FRANZ OPPACHER. Other intellectual debts are acknowledged through my frequent references to the bibliography.

I hope that the questions posed and the views advanced here may provoke the reader to further improve upon them.

Ottawa, November 1972

Charles Castonguay

Contents

Introduction

The referential view of meaning surely offers the most transparent conception of the meaning of an expression in a language, identifying as it does such meaning with the relation between the expression and the extralinguistic entities to which the expression is taken to refer. Whatever limited success this primitive view may have achieved in clarifying the meaning of concepts in factual theories dealing with the physical world, its application to mathematics immediately prompts delicate questions concerning the ontological and epistemological status of the presumed non-linguistic mathematical referents: *What is the mode of existence of these entities,* and *how does the mathematician have access to them?* In the present work we adopt an active mode of approach to these questions, and ask what sense can be *made* of mathematical existence, and how meaning and knowledge are *derived* in the course of mathematical activity.

To this end we will strive in what follows to draw close to what mathematics is, and attend to mathematics as process, rather than speculate on what it might be. And though a definite constructivist, kinetic view of mathematical certainty will emanate from our deliberations, we will not directly concern ourselves with the problem of the foundation of mathematical knowledge. Not that we feel that speculative philosophy lays any stronger claim on the latter problem; on the contrary, this is one philosophical domain where mathematics has seen to its own arms, and where significant discussion requires an especially developed mathematical competence, which we will not pretend to assume.

In spite of the silence of the various forms of mathematical Platonism, realism, and empiricism before the two questions which interest us, a woolly referential view of meaning is even today frequently advanced as providing an adequate explanation of the high measure of objectivity and consistency generally evidenced throughout the mathematical enterprise. The energies behind the logicist, formalist, and intuitionist drives towards a foundation for mathematical truth having been largely broken, spent, or rerouted, and having met, in any case, only with indifference in the mathematical community at large, the association between meaning and existence in mathematics appears to the contemporary working mathematician to remain as necessary as it is elusive.

Our main proposal is a dualistic approach to the problem of meaning, which, we feel, offers a more satisfactory treatment of meaning in factual contexts than does the pure referential view, and at the same time provides a validation of mathematical objectivity without evoking a vague ontology of mathematical objects. We regard meaning as an inseparable tissue of convention and fact. In seeking to lay bare the fine articulation of this tissue, we employ the hoary doctrine of intension and extension, which we interpret as basing meaning on the conjugation of linguistic relations of entailment with extralinguistic relations of reference. How we apply this inferential/referential framework in effecting a de-ontologization of mathematical meaning involves a fundamental transfer of the referential component of meaning to its heuristic component, briefly described below.

The notion of the extension of a concept, as the class of those objects to which the concept refers, is most often taken to be more transparent than that of intension, however sharply the medieval logicians conceived the distinction. Our proof-theoretic demystification of the intension of a concept, in terms of the class of those concepts which are logically derivable from the given concept, supplies a second definite pole, or axis, relative to which meaning may be conceived; a linguistic axis for meaning which proves to be, in a surprisingly precise sense, *dual* to that of reference. The flexibility of this dualistic view of meaning as comprised of intensional and extensional components articulated between such linguistic and referential (in technical logical terminology, "syntactical" and "semantical") poles, holds out sound promise as a tool for the elucidation of meaning in general, and, by laying sufficient stress on deduction, suggests an appropriate explanation of mathematical objectivity.

We do not, for all this, reject holus-bolus all talk of existence in mathematics. We endeavour instead, after scrutiny of some of the more careful treatments of the topic, to situate such talk in its proper place. As a result we avoid an overly narrow formalist attitude, and give a choice place, in our description of the process of mathematical enquiry, to what we call the *heuristic component* of mathematical meaning, as source of inspiration for the positing of relations between variously (and possibly referentially) perceived mathematical concepts or entities, relations which may eventually crystallize, through more exact formulation and deductive corroboration, into objective relations of entailment between linguistically expressed concepts.

This subsequently leads us to underline the essential contribution of style towards the development of a mathematical theory, in manifesting those significations which arise from a particular perception of the heuristic component of the theory. In attempting to circumscribe this phenomenon of mathematical style as heuristic catalyst, we find it necessary to distinguish between the meaning of a concept and what can be called its *signification,* style being the carrier of the more or less obscurely apprehended, relatively mute significations, while meanings are the conceptual, or linguistic, objectivizations of such significations.

Accordingly, the intensional component of meaning is delivered, for us, of the doubtful, murky, psychological quality which has made it so naturally suspect of subjectivity in the eyes of many. One might say that we choose to push back this murky dimension of meaning, to lay bare two objective components of meaning, one of which refers to objects, and which it is appropriate to name the extensional, or *correspondence,* component of meaning; the other relating to concepts or linguistic expressions, and which it is suitable to call the intensional, or *coherence,* component, in that it expresses how the given concept or expression coheres, or hangs together, with its fellows through relations of consequence.

The entire four chapters constitute a plea for the well-foundedness of this bifurcation, as introducing a minimal necessary complication to make up for the inadequacies of the early MORRIS-CARNAP acceptation of "semantics", as the simple relation of expressions to their designata. As we become engaged in the description of meaning in mathematics in particular, the importance of maintaining an objective linguistic pole for meaning becomes patent. For here the objectivity of factual referents, which serves as referential axis for meaning in physical theories or everyday discourse, gives way to multifaceted heuristic interpretations, none of which, quite often, may press any absolute claim to embody *the* semantics of a mathematical theory. Here objective meaning, first suggested through the different heuristic interpretations, is subsequently proven and stored in the coherence components, in intensions built upon inference. The entire heuristic component of a mathematical theory may be construed, in this way, as filling the void left by the absence of an authentic referential pole for meaning in mathematics, and as replacing existing objects in their contrapuntal role relative to intension.

Despite its reasonableness and practical appeal, such a view is not commonly given strong and clear voice in mathematics today, especially not in the preambles to contemporary introductions to mathematical logic. Not that a bold Platonism is flaunted there, either: the invitation to err is more subtle.

The position of set theory in contemporary mathematics is impressive, almost overwhelming. Largely due to the successful reduction to sets of the conceptual apparatus of all other branches of mathematics, and to the development in mathematical logic of adequate set-theoretic semantics for formalized theories, this preponderance weighs heavily on the philosophical problems before us. The temptation is strong to placidly accept as the "semantics" of a mathematical fragment, especially if the fragment is a formalized mathematical theory, a unidimensional set-theoretic interpretation of the fragment, rather than countenance a dynamic view of meaning as derivative of the imprecise but fundamental give-and-take between, on the one hand, entailment relations demonstrated within the fragment, and, on the other, less exact interconceptual relations intuited through the diverse interpretations making up the theory's heuristic component.

Certainly, despite the early paradoxes, the clarity and dependability of that portion of set theory required for the conceptual description of the vaster part of extant mathematics is seriously questioned by no one. However, we are not so much addressing ourselves to the problem of the foundation of mathematical knowledge as seeking to place in a dynamic perspective the impact of set theory on meaning and existence in mathematics. There are several reasons for not accepting set theory as furnishing a proper ground for an exclusively referential view of mathematical meaning, and for refusing set-theoretic reductions as offering a sufficient picture of mathematical semantics.

From the viewpoint of meaning-in-the-making, the value of set-theoretic reductions is seen to reside primarily in an improved facility for making the inter-theoretic comparisons and analogies upon which mathematical development thrives. In this sense, perhaps, a set-theoretic interpretation may provide an important contribution to the heuristic component of a theory. But meaning was alive and well in mathematics long before the advent of sets. Indeed, even as far as the question of foundations is concerned, we find no greater justification for epistemological pretensions on the part of set theory than there is for category theory: either offers a foundational style with enviable qualities of universality, suggestiveness, and consistency. To further maintain the absolute value of one foundational system over another is, to our taste, a futile exercise.

Consequently we find it undesirable to seat set theory on too lofty an epistemological pedestal. In particular, we see no reason in confusing objectivity with objects, and discount as epistemologically perverse the currently fashionable applications to mathematics of the Quinean doctrines of ontological commitment and ontological reduction. Our intent is by no means to generally disparage set-theoretic model theory as a possible formal elucidation of referential semantics. This notwithstanding, what seems to constitute the epistemological import of the classical reduction situations in mathematics is a minimal preservation, or isomorphism, of intensional meanings seconded by a certain *clash* of significations, resulting in a mutual interfecundation of the theories involved, and not in a passive reduction of one to the other.

The set-theoretic reduction of mathematics, the description of a mathematical theory in category-theoretic terms, the dualistic view of meaning as intension/extension, the heuristic appreciation of mathematical existence, all are as many styles expressing particular perspectives on their respective objects. In experiencing their epistemological appeal, we find a useful gauge in the distinction between process and product. What one philosophical view of mathematics advances may be convincingly applied to the finished product, but may fail to account for essential characteristics of the process of development; and *vice-versa*. A satisfactory epistemology of mathematics should straddle both aspects, and not, like logicism, early formalism, or Platonism, attend to foundations to the detriment of discovery, nor, through an overly fluid dialectic of reason and intuition, overemphasize process to the point of losing sight of objectivity. This being said, we find it philosophically more stimulating to attempt the problems posed by envisaging mathematics as a pursuit, or quest, perhaps only because of the general neglect of these problems. As a result we lay more stress here on the dynamic aspect of mathematics than is usual.

The formal exterior of the first chapter may belie this orientation. But behind its technical front it fulfills two basic goals. One is to demonstrate, through a careful explication of the traditional doctrine on intension and extension by means of the basic mechanisms of proof theory and model theory, the high validity of the claim that Tarskian set-theoretic semantics contains a satisfactory formal elucidation of the notion of reference. The other is to underline the possibility and the importance of cleaving to a linguistic, or syntactical, explication of entailment, as a necessary autonomous counterpoise to reference, without which the distinction between intension and extension becomes seriously degenerate.

Chapter 2 contains a discussion of coherence, correspondence, and dualistic views of meaning in general, and initiates our examination of the basis of mathematical objectivity. This leads to a first glance at meaning-preserving inter-theoretic correspondences, formalization, and the heuristic component of a mathematical theory. Model theory is seen as a higher-level intensional complement to proof-theoretic entailment in formalized theories, rather than as manifesting a properly referential aspect of mathematical meaning, patently absent in non-formalized theories. A comparison of our general view of meaning with that of C. I. LEWIS, an attempt at reconciling coherence and correspondence views of truth in mathematics, and a constructive criticism of two recent explications of the notion of nonexistent possibles, round out the chapter.

In the third chapter we cover considerable ground, perusing, to begin with, several conceptions of mathematical existence. The writings of BERNAYS and KREISEL, in particular, inspire our conclusion that what underlies the abundant and vague talk of existence in mathematics is the phenomenon of mathematical objectivity, rather than objects. Seating this objectivity on the evidence of mathematical proof, and observing that such objectivity is acquired, and that rigour is local, we strive to explain this relativization of mathematical objectivity by proposing a constructivist approach to the problem, based on PIAGET's genetic epistemology. PIAGET's vision of mathematics as progressing through successive reflective abstractions, beginning with the child's initial acquired awareness of the autonomy of sensory-motor operations with respect to the objects operated upon, growing through continually reviewed stages of conceptual re-elaboration, and terminating often enough in eventual formalization, illuminates the entire mathematical process of deriving meaning and acquiring objectivity. The explanatory power of this viewpoint is considerable, ranging in scope from the limitative results on formalization to the applicability of mathematics to reality, and rendering both of these phenomena equally comprehensible.

This view of objectivity as process, rather than as state, which shows the mathematician as deriving meaning not so much from objects, but from a sustained, precise kind of reflexion, leads us to answer the questions as to what positive sense can be made of speaking of existence in mathematics, and how such sense is made, through a fresh discussion of mathematical heuristics and realism. The diversity of stylistic approaches to a given mathematical subject, each of which, while reflecting a particular heuristic realization of the concepts in play, can in its own right prove fruitful, i. e., suggestive of confirmable theoretic consequences, attests at once to the anontological, logical nature of mathematical structures and to the mathematician's frequently rewarding practice of endowing them with ontological significations. We conclude that mathematical existence can properly be vindicated only heuristically. There ensues a brief account of the nature and function of style in mathematics, based on a recent essay by GRANGER, and of the interaction of style and concept in charting possible courses of action for the mathematical process.

The chapter closes with a renewed discussion of sets, categories, and mathematical semantics. We offer a brief description of the novel semantical basis for mathematics provided by category theory, underline the stylistic contrasts and possible areas of cross-fertilization between category theory and sets, and hint at some possible applications of categories to constructivist epistemology.

In the last chapter we propose an explication of the currently "hot" notion of inter-theoretic reduction, combining some past proof-theoretic work of KREISEL with our previous analyses of meaning and intension. This leads us to take a closer look at meaning-preserving inter-theoretic correspondences, following which we find it illuminating to distinguish between cases where reduction proper is at stake, and others which fall more under the general heading of explanation, a distinction largely meeting that found profitable earlier between static and dynamic aspects of mathematics. As dessert we try the ontological doctrines of QUINE.

QUINE is a notoriously successful smudger of distinctions. But his continuous spectrum of the sciences, running from ontology to mathematics and logic, leaves us uncomfortable and in doubt where existence and objective referential meaning are concerned. Somewhere along this spectrum, we feel that the significant ontological questions cross the theoretic/heuristic line, however vague that line may be. QUINE's style invites confusion of the formal, set-theoretic semantics of a formalized (or formalizable) mathematical theory with the "real", i.e., informal and heuristic, thing. Considering as we do the heuristic component to play much the same role, for a mathematical theory, that the referential component plays in a factual theory, we argue that the very number and nature of the so-called reductions in mathematics seriously, if not completely, depreciates the epistemological interest of the doctrine of ontological commitment, and renders ontological reduction, as QUINE applies it to mathematics, a twice hollow figure of speech.

1. Extension and Intension

1.1 The Basic Doctrine

While it is commonly held that the notions of extension and intension originate with the Port Royal logicians, a lively oral tradition has it that they are found explicitly in ARISTOTLE. In several passages of the *Organon,* ARISTOTLE does speak of the extension of the universals genus and species, often expounding, on the same occasion, the principle that the species is predicated of fewer things than the genus, and giving as example the genus "animal" and the species "man". He also asserts that the genus is predicated of the species, but not conversely. It can be reasonably argued that these assertions constitute the original formulation of the familiar relation between extension and intension, known as the law of inverse ratio: the extension of the genus contains the extension of the species, while the species comprehends or grasps more of the essence of a thing than does the genus.

FRISCH (1969) traces the development of the Aristotelian doctrine through PORPHYRY's *Isagoge* to the schoolmen, who transmitted it, though not without confusing changes in nomenclature, to the Port Royal era. In this respect, then, the Port Royalists' *La Logique ou l'art de penser* is not so much an isolated singularity in the history of logic as it was once thought to be.

But where the Port Royalists did hardily innovate is in breaking with the tradition of construing extension and intension only in terms of the genus and species, speaking instead of the extension (*étendue*) and the intension (*compréhension*) of ideas in general: "*J'appelle compréhension de l'idée, les attributs qu'elle enferme*

en soi, et qu'on ne lui peut ôter sans la détruire, comme la compré-
hension de l'idée du triangle enferme extension, figure, trois lignes,
trois angles, et l'égalité de ces trois angles à deux droits, etc. ...
J'appelle étendue de l'idée, les sujets à qui cette idée convient, ce
qu'on appelle aussi les inférieurs d'un terme général ..." [ARNAULD
and NICOLE (1662), p. 51].

In thus broadening the range of application of the notion of
extension, ARNAULD and NICOLE did not always employ an un-
ambiguous terminology. In particular, they do not state clearly
whether by *"inférieurs"* they understand species, or individuals, or
perhaps both[1]. That they did have in mind objects as *inférieurs,*
can be argued from their asserting that the extension of an idea
can be reduced *(resserrée)* without the idea being affected thereby
(op. cit., p. 52). Or again, were the relation whereupon their notion
of extension rests a relation between ideas, there would remain
little basis upon which to distinguish clearly between extension and
intension.

The Port Royal treatment of intension is also significantly original. The interconceptual relation which determines intension is
enlarged far beyond the simple relation of genus to species, to
encompass a more sophisticated logical organization of universals.
The inclusion, in the intension of the idea of a triangle, of the fact
that the sum of the angles of a triangle equals two right angles, involves the entire logical apparatus of Euclidean geometry in the
determination of intension. This opening of the traditional doctrine
to include more complex logical relations than mere predication in
the determination of intensions, suggests the following evaluation
of the Port Royal doctrine.

MOODY (1953) has brought to light in modern terms two basic
branches of medieval logic, the theory of supposition and the theory
of consequence. The latter is concerned with inference, or deduction. The former rests on the schoolmen's acute distinction between,
on the one hand, a relation of reference between universals and
things, according to which a universal may be said to be predicable
or not of a thing, and, on the other hand, an interconceptual rela-

---

1 See KNEALE and KNEALE (1962), p. 318. It does seem clear, however, that by *"sujet"* ARNAULD and NICOLE understand individuals, or
things: *"J'appelle chose ce que l'on conçoit comme subsistant par soi-*
même, et comme le sujet de tout ce que l'on y conçoit" (op. cit., p. 39).

tion between universals, according to which the genus is predicated of the species. The Port Royal definitions of extension and intension may be viewed as an attempt to mesh together the theories of supposition and consequence, by substituting more general methods of logical determination, taken from the theory of consequence, for the weak interconceptual relation of predication in the theory of supposition, and by generalizing the relation of reference in the latter theory to universals other than species.

It is difficult to begrudge the Port Royalists a few apparent inconsistencies in their presentation of such a daringly new perspective on logic[2]. We emphasize that this perspective calls for a comprehensive treatment of extension and intension, through the combination of the theories of supposition and of consequence — which would nowadays be called model theory and proof theory. This view does not encourage the division of logic into two mutually independent branches, extensional logic and intensional logic[3].

Once the breach in tradition was opened, the vocabulary of extension and intension became extremely varied. It is now common to speak of the extension and intension of terms and even of statements. The attributing of extensions and intensions to linguistic entities need not cause discomfort, however: linguistic entities can be considered to designate conceptual entities, so that while entertaining extensions and intensions at the linguistic level, it always remains possible, through this relation of designation, to reintegrate the world of concepts and ideas, if one so desires. We will often find it convenient to straddle both modes of speaking, in the sequel, by speaking of the extension and intension of *constructs,* where

---

2 Yet, notwithstanding the phenomenal advances in logic stimulated by the opening of this Pandora's box, FRISCH advocates a return to the Aristotelian doctrine; in his opinion, modern logicians, in detaching extension and comprehension from the predicables, have not, and could not, contribute anything worthwhile to the doctrine (op. cit., p. 127). The present chapter may be considered as arguing against such barren conservatism.

3 A division which LEIBNIZ, under the influence of DESCARTES, can be said to have overemphasized. It is perhaps due to the importance which LEIBNIZ attached to this division that the construal of extension in terms of individuals is sometimes erroneously considered as originating only with him (e. g. KAUPPI, 1967, p. 17). LEIBNIZ did originate, however, the present terminology of "extension" and "intension".

a construct can be envisaged as either a conceptual or a linguistic entity.

A variant notion of intension, in which dependence on a logical framework at the constructual level is abandoned in favour of a wholly referential approach, was distinguished from the traditional notion by KEYNES: "*Connotation* will then include only those attributes which are implied or signified by a name ... *comprehension* will include all the attributes possessed in common by all members of the class denoted by the name" [KEYNES (1887), p. 27]. The latter referential acceptation of intension constitutes a serious rupture in the tradition, as, under it, the extension of a construct would determine its intension, as we shall see below: this is contrary to the Port Royal conception.

In what follows, we will consistently employ *intension* when speaking of an adequate explication of the traditional doctrine taken in the broadest sense possible, reserving *comprehension* for an explication of the variant notion identified by KEYNES, based solely on the reference relation between constructs and objects; that is, comprehensions will be accepted as constituting intensions of a special kind[4].

1.2 A Set-theoretic Formulation

The basic doctrine on extension and intension can be stated very simply in the informal language of sets. All that is required is a set C of *constructs,* a set O of *objects,* a relation ϱ from C to O, and a relation η from C to C. The relation ϱ will be called a relation of *reference,* and may be viewed as an explication of that kind of relation which obtains when, in traditional terminology, a universal is predicable of a thing. The relation η of *entailment* can be viewed as representing that kind of relation which obtains when a universal is said to be predicable of another universal, or more broadly, as standing for a relation of consequence, or even, in certain referential contexts, as a relation of "comprehension" between constructs.

---

4 As we wish to examine what formal sense can be made of the doctrine of extension and intension, we will not discuss here the Aristotelian doctrines of essence and existence. Characters, properties, attributes, and the like purportedly serve to explain how the decision whether or not to predicate a universal of an object is to be made in practice.

Following standard usage, we write $c\varrho o$ to express the fact that the construct c is related by ϱ to the object o, and write $c\eta c'$ to indicate that η relates c to the construct c'; we say that c *refers,* or applies, to o, and that c *entails* or comprehends c'. We will say that the relations ϱ and η *constitute an explication of the notions of extension and intension* if and only if for any constructs c and c', if $c\eta c'$ then: (a) for any object o, if $c\varrho o$ then $c'\varrho o$, and (b) for any construct c'', if $c'\eta c''$ then $c\eta c''$. For a comparison with ARISTOTLE, c' may be taken here as a genus, and c as a species.

In our view, then, a relation of reference ϱ cannot alone explicate the notion of extension: it must, together with some relation η of entailment, satisfy the above desiderata. Any explication of extension must exhibit this minimal degree of articulation with an appropriate elucidation of intension. Likewise, no explication of intension is to be considered valid if unaccompanied by a suitable reference relation.

To give further transparency to our desiderata, we introduce the notation $\mathrm{Ext}\,(c)$ for the set of all objects to which the construct c is related by ϱ, and $\mathrm{Int}\,(c)$ for the set of all constructs to which c is related by η: that is,

$$\mathrm{Ext}\,(c) =_{\mathrm{df}} \{o\,\varepsilon\,O\,|\,c\varrho o\}$$

$$\mathrm{Int}\,(c) =_{\mathrm{df}} \{c'\,\varepsilon\,C\,|\,c\eta c'\}.$$

We will, of course, call $\mathrm{Ext}\,(c)$ the *extension* of c, and $\mathrm{Int}\,(c)$ the *intension* of c. Extensions of constructs are relative to the reference relation ϱ, while intensions are relative to the entailment relation η.

Our desiderata may now be rewritten in the following compact form:

(D 1) If $c'\varepsilon\,\mathrm{Int}\,(c)$, then $\mathrm{Ext}\,(c) \subseteq \mathrm{Ext}\,(c')$ and $\mathrm{Int}\,(c') \subseteq \mathrm{Int}\,(c)$.

This may be considered as the primitive, Aristotelian form of the law of inverse ratio of extension and intension. We state here the technically non-equivalent modern form:

(D 2) If $\mathrm{Int}\,(c') \subseteq \mathrm{Int}\,(c)$, then $\mathrm{Ext}\,(c) \subseteq \mathrm{Ext}\,(c')$.

As the relation of entailment is usually considered to be reflexive and transitive, the following proposition establishes that conditions (D1) and (D2) are generally equivalent:

Proposition 1: If η is reflexive and transitive, then (D1) and (D2) are equivalent.

Proof: Suppose the condition on η holds, and (D1) is satisfied. If $\text{Int}(c') \subseteq \text{Int}(c)$, since $c' \varepsilon \text{Int}(c')$ by reflexivity of η, we have $c' \varepsilon \text{Int}(c)$, whence $\text{Ext}(c) \subseteq \text{Ext}(c')$ by (D1). So (D2) is satisfied. Conversely, suppose (D2) is satisfied. If $c' \varepsilon \text{Int}(c)$, then by transitivity of η, $\text{Int}(c') \subseteq \text{Int}(c)$, whence further $\text{Ext}(c) \subseteq \text{Ext}(c')$ by (D2). So (D1) is satisfied.

In view of this equivalence, we will chiefly keep in mind (D2) as desideratum for explications of extension and intension, both because of its simplicity and because it immediately implies that

$$\text{if } \text{Int}(c) = \text{Int}(c'), \text{ then } \text{Ext}(c) = \text{Ext}(c'),$$

that is, equality of intensions forces equality of extensions.

It would be incorrect, however, to conclude from this that intensions "determine" extensions, and hence that extensions can be summarily dismissed as "dependent" upon intensions: the autonomy of the reference relation ϱ is only partly restricted by η under (D2). The whole interest of an explication of extension and intension lies, indeed, in how the two basic relations of reference and of entailment differ, while yet remaining linked together as in (D2). Thus under (D2) we do not have that entailment determines reference, nor that intensions determine extensions, but only that equality of intensions implies equality of extensions.

Essentially, then, (D2) requires only that a *dual homomorphism* of partially ordered sets be definable from the set $\mathscr{A}$ of all intensions of constructs, to the set $\mathscr{E}$ of all extensions, when both $\mathscr{A}$ and $\mathscr{E}$ are partially ordered by set inclusion[5].

---

5 A dual homomorphism between partially ordered sets P and Q is a function h from P to Q which reverses the order between its arguments: that is, if $a, b \, \varepsilon \, P$ and $a \leq b$ in P, then $h(b) \leq h(a)$ in Q. For the basic notions concerning ordered sets, lattices, and Boolean algebras which we shall utilize in this chapter, see BIRKHOFF (1966).

One expects more from a successful elucidation of extension and intension that the mere satisfaction of (D2). If we wish to extend the field of application of these notions to linguistic or conceptual forms other than the predicables, then we should determine, for instance, the relationship between $\text{Ext}(\sim c)$ and $\text{Ext}(c)$, where $\sim$ designates a unary operation of negation defined on the set of constructs C. In the next section we show how this can be done for any theory expressed in the language of classical first-order logic. But such considerations are recent, and it seems to us justifiable to consider that (D2) expresses most of what is clear and valuable in the old doctrines.

That the converse of (D2) be *false* — that the dual homomorphism pointed out above not be an isomorphism — could be added, in the spirit of Port Royal, as a further desideratum for a valid explication of intension and extension. For otherwise it would follow immediately that

$$\text{Int}(c) = \text{Int}(c') \text{ if and only if } \text{Ext}(c) = \text{Ext}(c')$$

which is counter to the explanations of the Port Royalists, as well as to the position of most modern exponents of the doctrine, for whom intensions must exhibit a certain primacy over extensions.

But rather than add this condition of asymmetry to (D2), we choose instead to consider formulations of intension which satisfy the converse of (D2) to be explications of a special kind. Let us say that an explication of intension is *proper* if it does not satisfy the converse of (D2), and that it is *referential* otherwise.

The notion of comprehension identified by KEYNES is an example of a referential explication of intension. Assuming as given a reference relation ϱ, the *comprehension* of a construct c, designated $\text{Com}(c)$, may be defined in our simple set-theoretic framework by

$$\text{Com}(c) =_{\text{df}} \{c' \varepsilon \, C \,|\, \text{Ext}(c) \subseteq \text{Ext}(c')\}.$$

Defining a relation η from C to C so that

$$c\,\eta\,c' \text{ if and only if } \text{Ext}(c) \subseteq \text{Ext}(c'),$$

one recognizes immediately that for this explication of entailment one has

$$\text{Int}(c) = \text{Com}(c).$$

That such η and ϱ do constitute a referential explication of intension (and extension) can be easily seen:

Proposition 2: $\mathrm{Com}\,(c') \subseteq \mathrm{Com}\,(c)$ if and only if $\mathrm{Ext}\,(c) \subseteq \mathrm{Ext}\,(c')$.

Proof: Suppose $\mathrm{Com}\,(c') \subseteq \mathrm{Com}\,(c)$. Since obviously $c'\varepsilon\,\mathrm{Com}\,(c')$, we have $c'\varepsilon\,\mathrm{Com}\,(c)$, whence $\mathrm{Ext}\,(c) \subseteq \mathrm{Ext}\,(c')$. Conversely, suppose $\mathrm{Ext}\,(c) \subseteq \mathrm{Ext}\,(c')$. If $c''\varepsilon\,\mathrm{Com}\,(c')$, this means $\mathrm{Ext}\,(c') \subseteq \mathrm{Ext}\,(c'')$, whence $\mathrm{Ext}\,(c) \subseteq \mathrm{Ext}\,(c'')$, that is, $c''\varepsilon\,\mathrm{Com}\,(c)$. Since c'' was arbitrary, $\mathrm{Com}\,(c') \subseteq \mathrm{Com}\,(c)$.

Let us call the relation of entailment η defined, as above, in terms of a given reference relation ϱ, the *entailment relation induced by ϱ*. It is now clear how the notion of comprehension, though originally defined only in terms of the relation of reference, furnishes nevertheless a referential explication of the notion of intension: namely, through the entailment relation induced by ϱ.

It is straightforward to check that the entailment relation comprised in any referential formulation of intension coincides with the relation of entailment induced by the corresponding reference relation. Comprehensions, then, provide the natural way of viewing intensions under a referential explication, as mirror-images of extensions: in such a case, the interdependence of reference and entailment is maximal.

1.3 Extension and Intension in Formalized Theories

We turn now to the problem of explicating extension and intension for constructs in formalized theories. By a theory we understand, intuitively speaking, any logically organized body of knowledge in which relations of inference, or entailment, or consequence, play a central role. We are aware that to speak of *formalized* theories is to commit oneself to a conflation of the theory being formalized, and the formalization of that theory in some formal language: this being said, for our immediate purposes we will accept this convenient *abus de langage,* though in later chapters it will be important to make the distinction[6].

---

6 We also will not preocuppy ourselves here with the distinction between the notions of an abstract language, and of a symbolic, or syntactic, representation of such a language. For such distinctions see CURRY and FEYS (1958).

Let us begin with the problem of explicating the notion of reference for such theories. There are no physical objects existing "out there" and to which the constructs of a formalized theory may be conceived to refer. Instead, in developing an explication of extension for such constructs, we will rely on the mathematical theory of models, initiated, in by far its greatest part, by the work of TARSKI (1956).

We will assume that we have at our disposal, therefore, the apparatus of interpreted formal calculi — interpreted, that is, in a certain set theory. For the sake of definiteness, we will focus on the notion of a model of a theory formalized in the language of classical first order logic (with equality); but in so far as the notion of a model is available for theories formalized in other formal languages, such as intuitionistic first order logic, or higher order languages, or infinitary languages, our definitions remain open to generalizations in these directions.

To fix a notation with which to discuss the elementary facts concerning set-theoretic models of first order theories, let us use P, Q, etc. for (possibly n-ary) predicate letters; x, x_1, x_2, y, etc. for individual variables; F, G, etc. for (well-formed) formulas; and the signs $\wedge$, $\vee$, $\sim$, $\rightarrow$, $\forall$, $\exists$, $($, $)$, for the usual logical symbols. All these constitute the symbolic apparatus of the formal language, the latter itself being designated by L. We will use T, T', etc. to designate theories expressed or formalized in L, and A, A', etc. for axioms of such theories. Finally, we use M, M', etc. to designate set-theoretic models for such theories; U, U', etc. for the universes of such structures; R, S, etc. for (appropriately n-ary) relations defined on the elements a, b, a_1, a_2, etc. of these universes (or of appropriate Cartesian products of them).

We recall that a *model* M for a theory T formalized in the language L consists of a universe set U, together with a set of relations on U indexed by the set of predicate letters of L in such a way that n-ary predicate letters are indices for n-ary relations, and such that every theorem F of T is true in M under the interpretation of the predicate letters defined by the given indexing. We assume familiarity with the notions of satisfaction and of truth (or validity) of a formula in a model.

In the process of defining the notion of truth, or satisfaction, a pair T, F or 0, 1 of "truth-values" is usually introduced *ex nihilo*, or else the universe U, taken from the structure under consideration, and the empty set ϕ are used for this purpose. Here we shall use as truth values the set ϕ and the set

$$O_M =_{df} \overset{\infty}{\underset{i=1}{\cup}} U^i$$

where U^i designates the ith Cartesian product of U. This settles the preliminary details required for our explication.

Given a theory T formalized in L, and a model M for T, we will consider the set of formulas of L as constituting a set of *constructs*, and the set O_M as constituting a set of *objects*, which we shall call *the set of objects taken from M*. We define a reference relation ϱ_M from constructs to objects, first by specifying that for any atomic formula $P\,(x_{i_1}, x_{i_2}, \ldots, x_{i_n})$, where all the x_{i_j}'s are distinct variables, we have

$$P\,(x_{i_1}, \ldots, x_{i_n})\ \varrho_M\ o \quad \text{if and only if } o\ \varepsilon\ R_P,$$

where R_P denotes that set in M indexed by P. Subscripting by M our notation for extensions in order to indicate relativity to the choice of model, we have immediately that

$$\text{Ext}_M\,[P\,(x_{i_1}, \ldots, x_{i_n})] = R_P.$$

For any open formula F we proceed similarly. If $x_{i_1}, x_{i_2}, \ldots, x_{i_n}$ are the distinct free variables of F listed in order of their occurrence from left to right in F, we stipulate that

$$F\,\varrho_M\,o \quad \text{if and only if } o \text{ satisfies } F \text{ in } M,$$

that is, if and only if $o = \langle a_1, a_2, \ldots, a_n \rangle\ \varepsilon\ U^n$ and $a_1, \ldots, a_n$ is an assigment of values to the variables $x_{i_1}, \ldots, x_{i_n}$ which satisfies F in M.

Finally, if E is a closed formula, i.e., a sentence, we stipulate that

$$E\,\varrho_M\,o \quad \text{if and only if } E \text{ is true in } M.$$

It follows immediately that if E is true in M then $\text{Ext}_M\,(E) = O_M$, and if E is false in M then $\text{Ext}_M\,(E) = \phi$.

Though the same universe may serve for different models M and M', in which case $O_M = O_{M'}$, as long as we are discussing extensions relative to one fixed model, we need not introduce some means of distinguishing between extensions of true sentences relative to such similar models, nor, for that matter, between extensions of false sentences, which according to our definition will always be the same set ϕ. A natural means of making such distinctions will suggest itself when we discuss at the same time extensions relative to several different models.

Care must be taken in visualizing the extension of the formula $P(x_1, x_2, x_2)$: we do not have $\text{Ext}_M[P(x_1, x_2, x_2)] = R_P$, since $R_P \subseteq U^3$ whereas $\text{Ext}_M[P(x_1, x_2, x_2)] \subseteq U^2$ according to our definition of ϱ_M. We also remark here that the extensions of two formulas which differ only in the variable letters occurring in them will be identical. And we do not have that for formulas F and G,

$$\text{Ext}_M[(F \wedge G)] = \text{Ext}_M(F) \cap \text{Ext}_M(G),$$

except when the list of distinct free variables of F is identical to that for G. For example, if F is $P(x)$ and G is $P(y)$, then $\text{Ext}_M[(F \wedge G)] \subseteq U^2$ whereas

$$\text{Ext}_M(F) \cap \text{Ext}_M(G) = \text{Ext}_M[P(x)] \cap \text{Ext}_M[P(y)] = R_P \cap R_P = R_p \subseteq U.$$

Thus the operation of taking set-theoretic extensions will only be compatible with the binary logical operations under suitable restrictions. More precisely, if Ext_M is considered as a mapping from the calculus of formulas to sets, then Ext_M is only a homomorphism from the calculi of *regular* formulas to the Boolean algebra of sets. We will define regular formulas and develop this last remark in Section 1.5.

If desired, one can also obtain a natural explication of extension for predicates, by taking the extension of the n-ary predicate letter P to be the extension of the formula $P(x_1, \ldots, x_n)$. But we will not deal with extensions and intensions of predicates here, as they can be subsumed in this way under extensions and intensions of formulas.

Traces of the vocabulary of extension in the study of formalized theories are sprinkled throughout the literature of modern logic since the time of FREGE. Our definition of the extension of a formula of an interpreted formalized theory is suggested by the well-worn, more or less explicit practice of discussing the extension, or "truth set", of a formula, or propositional function, found at elementary levels of mathematical activity as well as in advanced work (e. g., RUSSELL, (1919); HENKIN, (1950); HATCHER, (1968)). The use of model theory to explicate extension is quite natural: in fact, the elementary theory of models, itself, can perhaps best be described as resulting from the search for an explication of extension in formalized theories. This is one precise sense in which the parallel development of modern logic and of set theory can be described as a process of "extensionalization" of mathematics.

However, different model-theoretic explications of extension for formalized theories can be given. In particular, one could take the set of objects O_M to be the set of all infinite sequences of elements of U, and for any formula F, one could define $F_{\varrho M} o$ to hold for such an object if and only if the infinite sequence o presents an assignment of values to the infinite list of variables $x_1, x_2, \ldots$ which satisfies F. But under this more "Tarskian" formulation, $\text{Ext}_M (F)$ would consist of a set of infinite sequences of elements of U, no matter how many free variables occur in F, a prospect which makes our "truthset" definition appear more desirable: for one expects that two formulas, which differ in the number of distinct free variables occurring in them, will not refer to the same kinds of objects.

On the other hand, the Tarskian explication would be more sensitive to syntax than ours. For example, the extensions of two formulas which differ only in their variable letters would then not be identical. Even more, the mapping Ext_M would be a global homomorphism of syntactic structure onto a Boolean algebra of extensions (this approach is succinctly developed in HENKIN, 1955). The ontological ugliness of the Tarskian formulation is compensated for, in this way, by its syntactical sensitivity. But since we will have intensions at our disposal to handle syntactical matters, we can afford to allow extensions a greater degree of independence from syntax, and so we can prefer the more ontologically appealing truthset explication of extension.

While some attention has been given to the problem of explicating extension for formalized theories by way of model theory, the notion of intension, in contrast, appears to suffer from studied neglect in the metatheory of formalized theories. No doubt this is, in large part, due to the use of the adjective "intensional" to describe psychological situations in which extensions do not enjoy certain regulatory properties, as is the case, for example, with sentences expressing beliefs (cf. WHITEHEAD and RUSSELL, (1927), p. 73). Justifiable as such usage may be[7], starting from this, "intensional" unfortunately too easily becomes, for many logicians, a synonym for "messy" or "fuzzy".

---

7 Though this usage of the vocabulary of intension surely is also rooted in the Port Royal refusal of the converse of (D 2), we will not consider here such extensions of the basic doctrine on intension to pragmatics (in the sense of MORRIS-CARNAP).

Be that as it may, the entire metatheoretic apparatus required for a valid elucidation of intension in formalized theories is also already at hand: namely, proof theory and the explication it contains of the notion of entailment. In fact, we were overly hasty, above, in claiming a satisfactory explication for extension in terms of the notion of model alone, since in our view an explication of extension must be accompanied by an explication of intension, so that the validity of both may be checked according to the desiderata (D1) or (D2).

Assuming familiarity, then, with the notion of syntactical proof of a formula in a formalized theory, we define a relation η of entailment on the set of constructs (formulas) as follows. Let us say that a formula F is of *rank n*, and write rank$(F) = n$, if and only if there are in all n distinct free variables occurring in F. We use "$\vdash_T (F \rightarrow G)$" as shorthand for "the formula $(F \rightarrow G)$ is provable relative to the theory T". Given two constructs F and G, we now stipulate that

$$F \eta\, G \text{ if and only if rank}(F) = \text{rank}(G) \text{ and } \vdash_T (F \rightarrow G).$$

We have immediately that

$$\text{Int}(F) = \{G \mid \text{rank}(G) = \text{rank}(F) \text{ and } \vdash_T (F \rightarrow G)\}.$$

It is a trivial observation that this definition of η, together with a ϱ_M as defined earlier, establishes a valid explication of extension and intension according to desideratum (D2). For, given any model M of T, it is a consequence of the usual definition of satisfaction that, under the stated condition on free variables,

$$\text{if } \vdash_T (F \rightarrow G), \text{ then } \text{Ext}_M (F) \subseteq \text{Ext}_M (G).$$

This establishes (D2), since $\vdash_T (F \rightarrow G)$ is equivalent to $\text{Int}(G) \subseteq \text{Int}(F)$. In fact, since η is reflexive and transitive, our definitions satisfy (D1) also. It is obvious, too, that this explication is a proper (non-referential) one, in that a single ϱ_M will determine η only if the theory T is categorical.

Satisfaction of (D1) and (D2) far from exhaust the interrelations between extension and intension explicated in this way for formalized theories. We will see below that homomorphic calculi of extensions and intensions arise naturally from our explication. But for the moment, we show how a referential formulation of intension can also be given in the model-theoretic context.

1.4 Intension as Comprehension

By fixing on a model M of $\mathbf{T}$, and by taking the entailment relation to be the entailment relation induced by ϱ_M, as defined in the previous section, one immediately obtains a referential explication of intension (recall Proposition 2). Let us designate this entailment relation by η_M, and the corresponding comprehensions (or intensions) by $\mathrm{Int}_M\,(F)$: η_M would be an explication of entailment in only a very limited sense, as it would be satisfactory only for categorical theories.

Instead, let us consider the set $\mathcal{M}$ of all models of $\mathbf{T}$. For $M\,\varepsilon\,\mathcal{M}$, let O_M be the set of objects taken from M as described above, and let[8]

$$O =_{\mathrm{df}} \underset{M\,\varepsilon\,\mathcal{M}}{\cup}\ O_M.$$

For each $M\,\varepsilon\,\mathcal{M}$ we have a reference relation ϱ_M as defined above, and for each formula F we have an extension relative to ϱ_M such that $\mathrm{Ext}_M\,(F) \subseteq O_M$. Let

$$O_{\mathcal{M}} =_{\mathrm{df}} \mathcal{M} \times O,$$

the Cartesian product of $\mathcal{M}$ with O. Taking $O_{\mathcal{M}}$ as the set of objects, we define a relation $\varrho_{\mathcal{M}}$ between formulas and objects such that for any formula F we have

$$F\varrho_{\mathcal{M}} \langle M, o\rangle \text{ if and only if } o\,\varepsilon\,\mathrm{Ext}_M\,(F).$$

Let us designate the extensions taken relative to $\varrho_{\mathcal{M}}$ by $\mathrm{Ext}_{\mathcal{M}}\,(F)$; the induced entailment relation by $\eta_{\mathcal{M}}$; and the resulting comprehensions (intensions) by $\mathrm{Int}_{\mathcal{M}}\,(F)$. We note that, as promised earlier, extensions have been relativized to the different models in $\mathcal{M}$, in that objects taken from a given model are now tagged accordingly.

The entailment relation $\eta_{\mathcal{M}}$ so obtained coincides with the previously defined proof-theoretic entailment relation η. This is essen-

---

8 To call $\mathcal{M}$ a "set" is already illegitimate, to use it here as index "set" for a union operation is even more questionable. In VON NEUMANN-BERNAYS-GÖDEL set theory, $\mathcal{M}$ is generally a proper class, and not a set. We will overlook the set-theoretic technicalities here, as the difficulties connected with our constructions are not insurmountable. For instance, we may restrict our considerations to the set of all models in some sufficiently large universe of sets: then the set of models will simply not be a set in this universe.

tially the content of GÖDEL's Completeness Theorem for the predicate calculus. For suppose $F\eta_{\mathscr{M}} G$, i. e., suppose that $\text{Ext}_{\mathscr{M}} (F) \subseteq \text{Ext}_{\mathscr{M}} (G)$: this means that no matter what object $\langle M, o \rangle$ we choose, if $F\varrho_{\mathscr{M}} \langle M, o \rangle$ then $G\varrho_{\mathscr{M}} \langle M, o \rangle$, that is, if $o \varepsilon \text{Ext}_M (F)$ then $o \varepsilon \text{Ext}_M (G)$, or again, for all $M \varepsilon \mathscr{M}$, if the object o satisfies F in the model M, then o satisfies G in M. By GÖDEL's Theorem we infer that $\vdash_T (F \to G)$. If we have $\text{Ext}_{\mathscr{M}} (F) \neq \phi$, then we can conclude further from $\text{Ext}_{\mathscr{M}} (F) \subseteq \text{Ext}_{\mathscr{M}} (G)$ that $\text{rank}(F) = \text{rank}(G)$. However, if $\text{Ext}_{\mathscr{M}} (F) = \phi$, that is, if F is inconsistent with T, then for any G we have $\text{Ext}_{\mathscr{M}} (F) \subseteq \text{Ext}_{\mathscr{M}} (G)$: thus in order to ensure in such a case that $\text{rank}(F) = \text{rank}(G)$, we must add to the definition of $\eta_{\mathscr{M}}$ that it can hold only between formulas of equal rank. With this natural stipulation[9], we have that

$$\text{if } F\eta_{\mathscr{M}} G \text{ then } F\eta G,$$

and the converse follows from the definition of satisfaction. So for first-order theories, $\eta = \eta_{\mathscr{M}}$.

The pair of relations η, $\varrho_{\mathscr{M}}$ is, therefore, generally a proper explication of extension and intension, while the pair η, $\varrho_{\mathscr{M}}$ yields a referential explication. This suggests the following definitions. Let T be a (not necessarily formalized) theory, equipped with a syntax and a semantics, and containing an explication of the notion of entailment in its syntax. We may call T *semantically adequate* if its syntactical entailment relation coincides with the entailment relation induced from some reference relation defined in its semantics: otherwise, T is called *semantically inadequate*.[10]

In this sense any theory formalized in predicate logic is semantically adequate, by virtue of GÖDEL's theorem. As a similar com-

---

9 It is natural in that, under our truth-set explication of extension for formalized theories, the objects taken from a given model are of different kinds (as n-tuples of elements of the universe of the model), and so it is natural to construe a formula of rank n as only referring to, or talking about, objects of one fixed kind. Under the Tarskian explication of extension all objects are of the same kind (as infinite sequences of elements from the universe), and hence there would be no reason to exclude from $\text{Int}(F)$ or $\text{Int}_{\mathscr{M}}(F)$ formulas of rank different from that of F.

10 The adjectives "extensional" and "intensional" would most fittingly describe these situations, but we refrain from further compounding the already confused usage of these terms.

pleteness theorem has been shown to hold for higher-order languages, theories formalized in these languages also may be called semantically adequate. Intuitionistic, infinitary, and modal logics are also amenable to such description.

These results superficially suggest that in formal languages, the distinction between extension and intension, or between entailment and reference, is futile. On the contrary, the desire to achieve referential explications of intension has proven most fruitful: for example, HENKIN (1950) was led to consider non-standard models for higher-order logics, and KRIPKE (1965) invented a new type of model with respect to which intuitionistic logics are complete. The continuing dialectic between syntax and semantics is a principal axis of inquiry and inspiration in modern logic. As was already pointed out, that equality of intensions may, on certain occasions, be proven equivalent to equality of extensions, accomplishes no fundamental reduction of the notion of intension to that of extension, nor vice-versa: both are indispensable poles of language and thought. The problems of how entailment is determined, and of how reference is made, remain, in our view, as independent as are proof theory and model theory.

Whether or not a given theory is considered to be semantically adequate depends, of course, on what one accepts as constituting the semantics of the theory. In some formalized mathematical theories, such as predicate logic or group theory, it is natural to construe $\varrho_{\mathcal{M}}$ as the correct explicatum of reference within classical set-theoretic semantics, in view of the desirability of embracing a wide variety of non-isomorphic models for these theories. For other theories, such as arithmetic or the theory of functions of a complex variable, it is more natural to explicate reference by a relation ϱ_N for some "standard" or "intended" model N. The paradigm for the latter case is the usual interpretation given to GÖDEL's (First) Incompleteness Theorem, where N is taken to be the intended model of arithmetic (the set of natural numbers with $+$, $\cdot$, etc.): the theorem is interpreted as establishing that no recursively axiomatized theory formulated in predicate logic can capture all the truths of arithmetic, truth being taken relative to N.

Thus, GÖDEL's Incompleteness Theorem can paradoxically be read as asserting that any theory T which is axiomatized in predicate logic, and in which a certain minimum of arithmetic can be (syntactically) carried out, is semantically inadequate. For if G is a formula which is true in N but not provable in T, we have, for any logically true formula H with the same rank as G, that

$$\text{Ext}_N (H) \subseteq \text{Ext}_N (G) \text{ but Int}(G) \nsubseteq \text{Int}(H),$$

whence the pair η, ϱ_N constitutes a proper explication of extension and intension.

The paradox lies in that GÖDEL's Completeness Theorem was read as proving that any such **T** is semantically adequate. The paradox is dissolved by pointing out that in the one case we take ϱ_M as explicatum for reference, while in the other we take ϱ_N alone as explicatum, refusing to accept other ϱ_M's, with M not isomorphic to N, in our semantics. The Incompleteness Theorem is, in fact, often interpreted as proving that we must accept "non-standard" models in our semantics for such **T**'s, that is, as arguing for a broader semantics for formal number theories. The ambiguity here resides in that, within set-theoretic semantics, at least two distinct explications of reference are possible, hence two distinct "semantics" (even in the case of categorical theories, reference as ϱ_N is conceptually different from reference as ϱ_M).

GÖDEL's Incompleteness Theorem can also be described, in a natural way, as proving the *syntactical* inadequacy of certain theories, in that not all semantical truths of arithmetic can be captured syntactically. This further paradox, that the non-satisfaction of the converse of (D 2) can significantly be described as characterizing both syntactical and semantical inadequacy, can also be dismantled by adjusting one's perspective: a theory or language **T** for which the converse of (D 2) does not hold (relative to some fixed explication of reference) is syntactically inadequate in that there are relations (equations) of extension for **T** which have no intensional, or syntactical, counterpart; and is semantically inadequate for the same reason, only viewed differently — relations of extension cannot force corresponding relations of intension to hold.

The above discussion of GÖDEL's theorems illustrates the philosophical interest of recognizing the natural explications, in mathematical logic, of the traditional doctrine of extension and intension. These explications are also not without pedagogical value, in providing an intuitively appealing point of view from which to describe formal results in logic. But now let us turn from referential explications of intension, and give a more complete description of our proper explication.

1.5 Calculi of Extensions and Intensions

Given two predicates, for example "rational" and "animal", or "round" and "square", there is a long history to the discussion as to what extensions or intensions to attribute to the conjunction, disjunction, or other syntactical combinations of these constructs. We consider here such problems in the context of formalized theories, using the proper explication of extension and intension for such theories as developed, in Section 1.3, in terms of the pair η, ϱ_M for some fixed M.

As we remarked earlier, in order to ensure compatibility of the mapping Ext_M with the binary syntactical operations, we must restrict ourselves to special subsets of formulas. A formula F with exactly k free variables will be called *regular* if these variables, in order of their first free occurrence from left to right in F, form the list $x_1, x_2, \ldots x_k$. Let $\mathbf{F}^k$ designate the set of all regular formulas of L of rank k, for $k = 0, 1$, etc. Given a theory $\mathbf{T}$, let $\mathscr{F}^k$ designate the set of equivalence classes of $\mathbf{F}^k$ modulo the equivalence relation θ, where, for formulas F and G in $\mathbf{F}^k$, we have

$$F \theta G \text{ if and only if } \vdash_T (F \leftrightarrow G).$$

θ is compatible with the operations $\sim$, $\wedge$ and $\vee$, and $\mathscr{F}^k$ forms a Boolean algebra under the operations induced by $\sim$, $\wedge$ and $\vee$, which it is appropriate to call the *Lindenbaum algebra* of regular formulas of rank k.

If M is a model for $\mathbf{T}$ with universe U, we designate by Ext_M the mapping from the set of formulas into the power set $\mathscr{P}(O_M)$ which assigns to each formula its extension. Noting that for $F \, \varepsilon \, \mathbf{F}^k$ we have $\text{Ext}_M(F) \subseteq U^k$, and that Ext_M is compatible with θ in the sense that if $F \theta G$ then $\text{Ext}_M(F) = \text{Ext}_M(G)$, we have the following (where $U^0 =_{df} O_M$, and Ext_M designates ambiguously[11] the induced mapping on $\mathscr{F}^k$):

Proposition 3: For each k, Ext_M is a Boolean homomorphism from the Lindenbaum algebra $\mathscr{F}^k$ into the Boolean algebra $\mathscr{P}(U^k)$.

---

11 Strictly speaking, Ext_M is defined on formulas, not on equivalence classes of formulas, but it is convenient to employ the same designation for the mapping induced by Ext_M on the equivalence classes of formulas. This is permissible as long as the mapping one starts with is compatible with the equivalence relation in question.

Let $\mathscr{E}^k$ designate the set of extensions of all formulas in $\mathbf{F}^k$. Since $\mathscr{E}^k$ is the image of $\mathscr{F}^k$ under Ext_M, we can state:

Corollary: $\mathscr{E}^k$ is a Boolean subalgebra of $\mathscr{P}(U^k)$.

Proof: The above statements follow from the indicated compatibility conditions and from the obvious equalities resulting from the definition of satisfaction:

$$\mathrm{Ext}_M(\sim F) = -\mathrm{Ext}_M(F),$$

$$\mathrm{Ext}_M(F \wedge G) = \mathrm{Ext}_M(F) \cap \mathrm{Ext}_M(G),$$

and

$$\mathrm{Ext}_M(F \vee G) = \mathrm{Ext}_M(F) \cup \mathrm{Ext}_M(G),$$

where "$-$" designates set-theoretic complementation in $\mathscr{P}(U^k)$.

To match this calculus of extensions with a calculus of intensions, we could consider the set of intensions of regular formulas of rank k, designated $\mathscr{A}^k$, as also forming a Boolean algebra under the operations $\dot\sim$, $\dot\wedge$, and $\dot\vee$ defined so that

$$\dot\sim \mathrm{Int}(F) =_{\mathrm{df}} \mathrm{Int}(\sim F),$$

$$\mathrm{Int}(F) \dot\wedge \mathrm{Int}(G) =_{\mathrm{df}} \mathrm{Int}(F \wedge G),$$

and

$$\mathrm{Int}(F) \dot\vee \mathrm{Int}(G) =_{\mathrm{df}} \mathrm{Int}(F \vee G).$$

Designating by Int the mapping from $\mathbf{F}^k$ to $\mathscr{P}(\mathbf{F}^k)$ which assigns to each formula its intension, noticing similar compatibility conditions as before, and subscribing to an ambiguous convention of designation similar to that used for Ext_M above, Int could be viewed as a Boolean isomorphism (since $\mathrm{Int}(F) = \mathrm{Int}(G)$ if and only if $F \theta G$) from the Lindenbaum algebra $\mathscr{F}^k$ to the Boolean algebra of intensions $\mathscr{A}^k$. But while Proposition 3 was arrived at naturally, such an arbitrarily defined calculus of intensions is not well-motivated.

Our distaste for a calculus of intensions defined in this way finds natural grounds in the fact, already pointed out in Section 1.2, that $\mathscr{A}^k$, as a set partially ordered by set inclusion, is *dual* homomorphic to the similarly ordered set $\mathscr{E}^k$. Moreover, if T is a categorical theory, or if we are dealing with intensions referentially defined (i. e., with comprehensions), then the Boolean calculus of extensions induces, in a natural way, a *dual* Boolean calculus of intensions.

To better see this latter point, let us use simply Com(F) and Ext(F) to designate intensions and extensions in such a case. Then Com($F \wedge G$) is a least *upper* bound for Com(F) and Com(G) in $\mathscr{A}^k$. For since Ext($F \wedge G$) = Ext(F) $\cap$ Ext(G) $\subseteq$ Ext(F), by Proposition 2 we have Com(F) $\subseteq$ Com($F \wedge G$); similarly one obtains Com(G) $\subseteq$ Com($F \wedge G$) and so Com($F \wedge G$) is an upper bound for Com(F) and Com(G). Next suppose that H is such that Com(F) $\subseteq$ Com(H) and Com(G) $\subseteq$ Com(H); then by Proposition 2, Ext(H) $\subseteq$ Ext(F) and Ext(H) $\subseteq$ Ext(G), whence Ext(H) $\subseteq$ Ext(F) $\cap$ Ext(G) = Ext($F \wedge G$), and by Proposition 2 again, Com($F \wedge G$) $\subseteq$ Com(H), i. e., Com($F \wedge G$) is a least upper bound. It is just as easy to show that Com($F \vee G$) is a greatest lower bound for Com(F) and Com(G) in $\mathscr{A}^k$, and that the remaining properties of a complemented distributive lattice can be induced from the Boolean properties of $\mathscr{E}^k$. The Boolean algebras $\mathscr{A}^k$ and $\mathscr{E}^k$ are in fact dual isomorphic in this case, by Proposition 2.

Therefore we reject our earlier definitions of $\wedge$ and $\dot{\vee}$ [12] in favour of the dualized definitions

$$\text{Int}(F) \wedge \text{Int}(G) =_{df} \text{Int}(F \vee G)$$

and

$$\text{Int}(F) \dot{\vee} \text{Int}(G) =_{df} \text{Int}(F \wedge G),$$

the definition of $\sim$ remaining unchanged. Adopting the usual ambiguous convention of designation for Int, we have:

Proposition 4: For each k, $\mathscr{A}^k$ is a Boolean algebra, and Int is a dual Boolean isomorphism from the Lindenbaum algebra $\mathscr{F}^k$ onto $\mathscr{A}^k$.

In fact, since considerations of rank and regularity are of no importance here (extensions are not involved), this result can be extended to a full calculus of intensions $\mathscr{A}$ and to the full Lindenbaum algebra $\mathscr{F}$.

Unlike the algebra $\mathscr{E}^k$, which is a Boolean subalgebra of $\mathscr{P}(U^k)$, the algebra of intensions is not compatible with the set-theoretic Boolean operations on $\mathscr{P}(F^k)$. For instance, it is trivial to find examples where Int($\sim F$) $\neq -\text{Int}(F)$. All that can be asserted in this direction is that Int(F) $\wedge$ Int(G) = Int(F) $\cap$ Int(G), and that Int(F) $\cup$ Int(G) $\subseteq$ Int(F) $\dot{\vee}$ Int(G) [13]. That the latter inclusion may be

---

12 Though they are proposed, as they stand, for the calculus of intensions by SUSZKO (1967), p. 21.

13 One might find grounds in this incompatibility for calling the calculus of intensions "intensional".

strict can be seen from the well-worn illustration from Euclidean geometry, that neither "equilateral" nor "triangular", taken separately, entails "equiangular", whereas their conjunction does.

Applying first to the set Int(F) the inverse of the isomorphism Int of Proposition 4, then following up with the homomorphism Ext$_M$ from Proposition 3, we obtain;:

Proposition 5: For each k, a dual homomorphism of Boolean algebras from $\mathscr{A}^k$ onto $\mathscr{E}^k$ is defined by mapping Int(F) onto Ext$_M$ (F).

This dual homomorphism is a dual isomorphism under a referential explication of extension and intension.

The set of logical operations under consideration can be extended to include existential and universal quantification by noticing that the quantifiers give rise to mappings from $\mathscr{E}^k$ to $\mathscr{E}^{k-1}$ and from $\mathscr{F}^k$ to $\mathscr{F}^{k-1}$. For example, let $\pi_k \colon U^k \to U^{k-1}$ be the projection mapping such that for $\langle a_1, a_2, \ldots, a_k \rangle \ \varepsilon \ U^k$,

$$\pi_k \left(\langle a_1, \ldots, a_k \rangle \right) = \langle a_1, \ldots, a_{k-1} \rangle \ \varepsilon \ U^{k-1}.$$

We also have, at the syntactical level, the mapping

$$(\exists \ x_k) : \mathscr{F}^k \to \mathscr{F}^{k-1}.$$

Using the definition of satisfaction one can show, for any formula F, that

$$\text{Ext}_M \left[(\exists \ x_k) \ (F) \right] = \pi_k \left[\text{Ext}_M \ (F) \right].$$

Briefly, $\mathscr{P} \ (U^k)$ and $\mathscr{F}^k$ can be viewed as *cylindric* algebras (cf. HENKIN, MONK, and TARSKI (1971)), and Ext$_M$ becomes a homomorphism of cylindric algebras. There is no difficulty in giving $\mathscr{A}^k$ a cylindric structure also, and Int becomes an isomorphism of cylindric algebras. Propositions 3 to 5 carry through: but we omit the details [14].

---

14 The rank and regularity restrictions can be removed from the above propositions if the Tarskian explication of extension is adopted. This results in a smoother global articulation between $\mathscr{A}$ and $\mathscr{E}$, which is no surprise, as we remarked earlier that the Tarskian definition of extension is more sensitive to syntax than is the truth-set explication. In any case, under either acceptation of extension, rank and regularity conditions can be removed from Proposition 4, and a natural and global calculus of intensions, dual isomorphic to the full Lindenbaum calculus, can thus be obtained.

Our description of the relation between the syntax and semantics of formalized theories is now sufficiently complete to permit us to conclude that the primitive duality between entailment and reference, as conceived traditionally and resumed in (D2), has been most successfully developed within modern logic; to the point that a formalized theory can boast, together with a calculus of extensions, a dually articulated calculus of intensions. We now turn to some other facets of the formal elucidation of extension and intension which will be relevant later on.

1.6 Extension and Intension of Theories

It is natural to define an entailment relation, which we again designate by η, between sets of formulas of a language L, such that if F and G designate sets of formulas, we have

$$F \, \eta \, G \text{ if and only if } F \vdash_L H \text{ for all } H \, \varepsilon \, G.$$

It is also natural to define a reference relation ϱ_M or $\varrho_{\mathcal{M}}$ for sets of formulas, either relative to a fixed model M of L or to the set $\mathcal{M}$ of all models of L. For ϱ_M one can take $\{M\}$ for the set of objects, stipulating that

$$F \varrho_M M \text{ if and only if } M \text{ models } F;$$

for $\varrho_{\mathcal{M}}$ one can take $O = \mathcal{M}$, again with

$$F \varrho_{\mathcal{M}} M \text{ if and only if } M \text{ models } F.$$

It is obvious that either ϱ_M or $\varrho_{\mathcal{M}}$, combined with η, yields an explication of extension and intension for sets of formulas, generally a proper explication with ϱ_M, and a referential one with $\varrho_{\mathcal{M}}$. To a certain extent, the set-theoretic calculus of sets of formulas is dually reflected by both reference relations into the set-theoretic calculus $\mathscr{P}(O)$, for example $\text{Ext}(F \cup G) = \text{Ext}(F) \cap \text{Ext}(G)$ in both cases. However, complementation does not carry through: for instance, $\text{Ext}(F) = \text{Ext}(-F) = \phi$ under both explications of reference, if both F and $-F$ are inconsistent sets of formulas[15].

---

15 A more sensitive calculus of extensions than the bivalent "truth-value" calculus can be defined relative to a fixed M, using the Tarskian explication of extension. The set of objects is taken to be the set of all

A theory T formalized in L is a set of formulas which is closed under η, in the sense that if T η F, then F $\subseteq$ T. Just as it was esteemed more natural, in Section 1.3, to explicate entailment for formulas in such a way that a formula F can only entail formulas of the same rank as F, so is it proper, when discussing entailment among theories, to consider only those sets of formulas which are themselves theories, as possibly being entailed by a given theory. We continue to use η to designate this more restricted entailment relation. Reference for theories can be explicated either by ϱ_M or $\varrho_\mathscr{M}$, as defined above for sets of formulas.

If T and T' are theories, by the closure of T we have immediately that

$$\text{T } \eta \text{ T' if and only if T'} \subseteq \text{T}.$$

Hence the *intension* of a formalized theory T may be simply defined to be the set

$$\text{Int(T)} =_{\text{df}} \{\text{T'} | \text{T'} \subseteq \text{T}\},$$

that is, the set of all its subtheories. The *extension* of T may be taken as $\{M\}$ if M models T, and ϕ otherwise, if reference is explicated by ϱ_M; or as

$$\text{Ext}_\mathscr{M} \text{ (T)} =_{\text{df}} \{M \ \varepsilon \ \mathscr{M} | M \text{ models T}\},$$

if $\varrho_\mathscr{M}$ explicates reference. Generally, the pair η, ϱ_M is a proper explication of extension and intension for formalized theories, and the pair η, $\varrho_\mathscr{M}$ is a referential explication if L is semantically adequate.

TARSKI's calculus of formalized theories ((1956): he calls them *formal systems*) could be used to obtain calculi of intensions and extensions for theories. TARSKI defines operations $\cdot$, $\dotplus$, and $^-$ on theories by

$$\overline{\text{T}} =_{\text{df}} \bigcap_{F \ \varepsilon \ \text{T}} \text{Cn} (\{\sim F\}),$$

$$\text{T} \dotplus \text{T'} =_{\text{df}} \text{Cn} (\text{T} \cup \text{T'}),$$

and

$$\text{T} \cdot \text{T'} =_{\text{df}} \text{Cn} (\text{T} \cap \text{T'}),$$

---

sequences of elements of the universe of M, and one defines ϱ_M by $F \varrho_M o$ if and only if the sequence o satisfies all the formulas in F. The resulting calculus is nonetheless as imperfect as those obtained above.

where $Cn(G) =_{df} \{H | G \vdash_L H\}$ [16]. With set inclusion acting as a partial order between theories, the calculus obtained has all the properties of a Boolean algebra, except that the law of excluded middle is not satisfied, that is, $T + \overline{T}$ is not always the entire set of formulas of L (this is due to the difficulty in defining the notion of complement of a theory).

Thus, with TARSKI's calculus playing the role of the Lindenbaum algebra, formal calculi of extensions and intensions and homomorphism theorems could be obtained as in the preceding section. But the calculi are generally not Boolean, and the homomorphisms lack all but formal interest, since (dual) compatibility of the defined operations with the set-theoretical operations already escaped us in the calculus of extensions of theories. For one easily finds that $Ext(T) \cap Ext(T') = Ext(T + T')$, but one can only assert that $Ext(T) \cup Ext(T') \subseteq Ext(T \cdot T')$ and $Ext(\overline{T}) \subseteq - Ext(T)$, where extensions are defined relative to either ϱ_M or $\varrho_{\mathcal{M}}$.

Only for the subset of finitely axiomatizable theories does one obtain results as complete as those of Section 1.5. Only for such theories are the calculi Boolean (TARSKI, op. cit., p. 355), and it is a simple matter to also check (dual) compatibility of the calculus of extensions with the set calculus of $\mathscr{P}(O)$. We note that the Tarskian calculus of theories does not quite play, relative to theories, exactly the same role as does the Lindenbaum algebra relative to equivalence classes of formulas, since the compatibility results are dualized, in that suprema are mapped by Ext into infima, and *vice-versa*. But all becomes clear when one recalls that the calculus of finitely axiomatizable theories is *dual* isomorphic to the Lindenbaum algebra (ibid., p. 355). It follows that the calculus of finitely axiomatizable theories is isomorphic to the calculus of intensions of formulas in L.

We sum up these results concisely by momentarily adopting the following ambiguous conventions: $\mathscr{A}$ will designate the calculi of intensions of formulas or of theories, $\mathscr{F}$ the Lindenbaum algebra of equivalence classes of formulas or the Tarskian calculus of theories, $\mathscr{E}$ the calculi of extensions of formulas or theories, and $\mathscr{P}$ the set-theoretic calculi of objects (sets of objects taken from a model, or sets of models); "$\rightarrow$" will indicate homomorphism, "$\nearrow$" a dual

---

16   Cn (G) is not Int (G) in the terminology of the beginning of this section, since Int (G) is a set of *sets* of formulas: Cn (G) is, precisely, the greatest set in Int (G).

homomorphism, "$\xrightarrow{e}$" compatibility, and the two-headed arrows indicate isomorphism. For formulas one has the following diagram:

$$\mathscr{A} \diagup \mathscr{F} \rightarrow \mathscr{E} \xrightarrow{e} \mathscr{P}$$

for theories in general one will have:

$$\mathscr{A} \leftrightarrow \mathscr{F} \diagup \mathscr{E}$$

and for finitely axiomatizable theories one obtains:

$$\mathscr{A} \leftrightarrow \mathscr{F} \diagup \mathscr{E} \xrightarrow{e} \mathscr{P}.$$

One can link together the first and last diagrams with suitable arrows as mentioned above.

1.7 Intension as Connotation: Core Intension

The intension of a construct c is often conceived as a finite set of constructs which "make up the meaning" of c, or as a finite set of characteristics "necessarily" implied by c. A modal formulation of necessity is not a prerequisite for the explication of intension in the latter sense. When KEYNES agrees with J. S. MILL that "equiangular" is not part of the intension of "equilateral and triangular" (KEYNES, op. cit., p. 25), we may simply take it that KEYNES and MILL are not assuming all of Euclidean geometry as part of the background theory T relative to which intensions are determined: what is "necessary" is relative to T.

Intension as a finite set can be explicated through the notion of core intension (our usage of this term differs from that of BUNGE, 1967). Assuming that an explication of extension and intension has been given, a *core intension* of c is any finite subset D of Int(c) such that D η Int(c), that is, such that Cn(D) = Int(c). A core intension D of c may be construed as giving the "surface" intension of c, whereas Int(c) gives the intension of c "in depth", using the full logical apparatus of the language or theory.

Extending the definition to sets of formulas, a core intension for a theory T would, in particular, be nothing else than a finite set of axioms for T. It follows that not all sets of formulas boast core intensions (whereas all formulas do).

The notion of a core intension brings to mind the usage of a dictionary in seeking out surface meanings. Core intensions may be thought of as conventionally determined by means of a "dictionary" function, designated Dic, from C to $\mathscr{P}(C)$, which assigns to each construct c a finite set of constructs, designated Dic(c), such that Dic(c) η Int(c).

Indeed, Dic itself may be used as a first approximation of entailment, say for very primitive languages. We can define recursively what we shall call the *connotation* of c, designated Con(c), by

$$\mathrm{Con}(c) =_{\mathrm{df}} \bigcup_{c' \,\varepsilon\, \mathrm{Dic}\,(c)} \mathrm{Con}(c').$$

Such an explication, which corresponds to the naive idea of taking the meaning of a construct to be what one eventually can find through consulting a dictionary, would seem better suited to KEYNES' and MILL's notion of intension ("connotation" in KEYNES' terminology — see the quotation from KEYNES in Section 1.1) than our definitions of intension in Section 1.3 or 1.4. The primitive conceptual framework provided by Con(c) and Dic can be considered to be an ancestor of the sophisticated clarifications of intension and entailment provided by modern logic.

1.8 Vagueness

When explicated formally in set-theoretic semantics, as above, reference benefits from the law of excluded middle, a standing rule of classical mathematics: by the definition of the notion of satisfaction, for any formula F, either F or $\sim F$ will apply to any one object. If extension of our above analysis to the explication of extension and intension in factual theories or natural languages is envisaged, referential and linguistic vagueness must be taken into account. We will assume here that the language in question has operations which (vaguely) correspond to negation, disjunction and conjunction.

Extensional vagueness, that is, the difficulty of recognizing an object as being referred to by a word, forces the abandoning of the law of excluded middle, in favour of Ext($\sim F$) $\subseteq$ $-$Ext(F), since there may now be objects which, for subjective, perceptual, definitory, or other reasons, it is impossible to consider as objects to which either F or $\sim F$ applies. Let us call an object an *F-object* if F applies to it. We can expect that vague disjunction will, nevertheless, be sufficiently exact to guarantee that any F-object will be an $(F \lor G)$-object, that is, that

$$\text{Ext}(F) \cup \text{Ext}(G) \subseteq \text{Ext}(F \lor G).$$

But extensional vagueness works against equality here too: for example, one may hesitate to call an object "pink", or to call it "red", yet readily affirm that it is definitely "either pink or red". So we cannot assert the converse inclusion.

Intensional vagueness[17], or the difficulty of precisely delimiting the effect, on intensions, of the linguistic operations, may be at work in this last inclusion also, giving rise to a lenient, permissive attitude in determining the extension of a disjunction in terms of its disjuncts. Intensional vagueness, much more than extensional, furnishes natural motivation for adopting the inclusion dual to that above, namely,

$$\text{Ext}(F \land G) \subseteq \text{Ext}(F) \cap \text{Ext}(G):$$

a conservative attitude in the use of (vague) conjunction may result in one's seeing fit to consider an object separately as an F-object, and as a G-object, but not as an $(F \land G)$-object. More precisely, the intensional effect of conjunction is generally multiplicative rather than additive — to wit, the generally proper classical inclusion $\text{Int}(F) \cup \text{Int}(G) \subseteq \text{Int}(F \land G)$ pointed out in Section 1.5.

If one retains certain natural consistency assumptions, a lattice of extensions nevertheless can be obtained. Such a consistency assumption is already embodied in $\text{Ext}(\sim F) \subseteq -\text{Ext}(F)$, but one can also argue for consistency in the use of conjunction and disjunction. For instance, it is natural to assume that if every F-object is an H-object, and if every G-object is an H-object, then every $(F \lor G)$-object is also an H-object. Combined with $\text{Ext}(F) \cup \text{Ext}(G) \subseteq \text{Ext}(F \lor G)$, this guarantees that $\text{Ext}(F \lor G)$ is a least upper bound for $\text{Ext}(F)$ and $\text{Ext}(G)$. A similarly motivated consistency assumption guarantees that $\text{Ext}(F \land G)$ is a greatest lower bound for $\text{Ext}(F)$ and $\text{Ext}(G)$. Dual assumptions such as $\text{Int}(F) \cup \text{Int}(G) \subseteq \text{Int}(F \land G)$ are natural, and a dual lattice of intensions can be obtained. But neither lattice is

---

17 CARNAP (1955) improperly refers to extensional vagueness as "intensional" vagueness. But then, CARNAP does not hesitate, either, to explicitly conflate referential and proper explications of intension (cf. CARNAP, 1956, p. 67).

compatible with the set-theoretic operations of their underlying sets, and neither is complemented.

To assume more structure on the part of vague intensions and extensions would require deeper pragmatic analysis than we wish to enter into here. One promising direction which such inquiry could take would be to investigate the correlations between vagueness and intuitionistic logic. Instead, we comment briefly on two recent approaches to the problem.

GOGUEN (1969) bases his explication of inexactness on extensional considerations only. His method is to increase the range of truth-values which can be assigned to the application of a predicate to an object, by using a complete-lattice-with-an-additional-semigroup-operation, called a "closg", as a system of truth-values. He often uses the unit interval [0, 1], ordered as usual, as an example of a closg, with the supremum of two numbers defined to be the largest of the two, the infimum defined as the smallest, and with multiplication as the additional semigroup operation (designated by $\cdot$). Designating by $[F]$ the truth-value assigned to the applicability of a predicate F to an object, GOGUEN considers that $[F \wedge G]$ should generally be less than $[F] \wedge [G]$ in the [0, 1] closg, and suggests taking $[F \wedge G] = [F] \cdot [G]$. The motivation agrees with ours when we proposed the inclusion $\text{Ext}(F \wedge G) \subseteq \text{Ext}(F) \cap \text{Ext}(G)$ in vague contexts.

However, under GOGUEN's proposal one would always have that $[F \wedge F]$ is strictly smaller than F, and often intolerably smaller; for example, if $[F]$ is 0.1 for a given object, then $[F \wedge F]$ would be 0.01, which is absurd. Not only does GOGUEN ignore the obvious desiderata that $[F \wedge F]$ and $[F \vee F]$ should be relatively close to $[F]$, but on the basis of his closgian semantics he proposes "truth set rules" where the closg operations all figure in the syntax of his inexact languages (ibid., p. 362), without their having been given any intrinsic syntactical sense. When one thus seeks to explicate vagueness by extensional vagueness alone, the explication runs the risk of making no sense at the intensional (syntactical) level.

Another fault which we find in GOGUEN's approach, is the assumption that for *all* objects, and for *all* formulas, $[F \wedge G]$, $[F \vee G]$ and $[\sim F]$ can be calculated in the *same* way from $[F]$ and $[G]$ using the operations of the one underlying closg. We see no reason to accept such a vast assumption of uniformity, at least not where natural languages are concerned. Again, because of his unfortunate choice of multiplication on [0, 1] as the semigroup operation, GOGUEN finds that the naturally defined closgian pseudo-complement of every number in [0, 1] is 0, and so decides that it would be best to consider closgs-with-negation, the negation operation being practically independent of the closg structure. This move is made even more suspect by GOGUEN's choice of the "familiar complement ... which assigns 1-a to a" as model for negation on [0, 1], and by his finding that "for many applications [none of which is mentioned] this complement is a better model of negation than the pseudo-complement" (ibid., p. 359).

This assumption of the absolute coherence of semantical fields, and this explication of negation as an *exact* operation in the logic of *inexact* concepts, do not encourage us to expect that by pursuing investigations in this direction "a satisfactory account of semantics in natural languages may eventually be possible" (ibid., p. 336).

KÖRNER (1960) defines a concept to be *inexact* "if both its assignment and its refusal to some object would conform to the rules [of reference] governing it" (ibid., p. 161); *exact* concepts are those for which there is no such ambiguity. KÖRNER's "inexact" rules of reference would more properly be called "incomplete", in that, as far as we can gather, their inexactness does not depend on any intensional vagueness, nor on their being inconsistent, but on their being insufficiently determinate: one person may (consistently) apply an inexact concept to a given object, while another person may (equally consistently) refuse the concept to the same object.

Rather than discuss KÖRNER's notion of exactness (for such debate see KUMAR, 1967, and KÖRNER, 1967), we will consider the application he makes of it to his central problem, that of how mathematics is applied to reality. He asserts that the concepts of mathematics are "purely exact", and explains that applications of mathematics to reality occur through the interchange of exact (mathematical) and inexact (physical) concepts.

Though KÖRNER emphasizes that he intends his definition of exactness to be ontologically neutral, in that whether or not a concept is exact does not depend on which objects are "real" and which objects are not ((1960), pp. 160 ff.), this is no warrant, in our opinion, for the applicability of his definition to mathematical concepts: we argue in Chapter 3 that mathematics is not only ontologically neutral, but is in fact ontologically null.

Indeed, the only justification which KÖRNER gives for considering mathematics as exact is not an ontology, or theory of reference for mathematics, through which we might visualize his basic definition in terms of certain "objects", but the fact that "in the last resort all mathematics can be presented in terms of two notions, that of a set or range of an *exact* concept (propositional function, etc.) and that of a function (mapping, etc.) defined in terms of 'set'" (ibid., p. 168). To thus rest the exactness of mathematics, as defined by KÖRNER, on that of the notion of set, not only opens wide the doors to all manner of ontological reductions (an attitude which we also criticize in Chapter 4), but is cogently refutable by the independence results in set theory, which may be interpreted as establishing the vagueness of the notion of set.

To proclaim mathematics exact in KÖRNER's ontological sense is, we repeat, make-believe. Certainly, it is a commonplace that mathematical concepts are exact: but, as we shall also argue in Chapter 3, exactness is in fact an objective in mathematics, not a reality[18].

1.9 Intensional Autonomy

The valid explications of extension and intension in formalized theories which we have given in Section 1.3, and developed in 1.5, seem too obvious to dwell upon. Yet, all of the explications of these notions offered thus far in the literature have treated intension referentially, in terms of the class of all models of a theory, instead of explicating intension linguistically, at the syntactical level, which we find is more proper. Accompanying such a referential approach to intensions is the failure to formulate any desiderata such as (D 1) or (D 2), with respect to which the proposed formal elucidations may be validated as explicating the traditional doctrine.

-----

18 This applies as well to the notion of proof, which KÖRNER also takes to be exact (ibid., p. 161).

Whether to take mathematical concepts as exact or not hardly seems to make much difference to KÖRNER's explanation, anyway: "That satisfiable rules governing — more or less strictly — the interchange of exact and inexact concepts (before and after mathematical deduction) have been found, depends on those features of the world which go under the name of human ingenuity" (ibid., p. 180). Careful case studies of some of the more ingenious among these rules of interchange of mathematical and factual concepts would be more enlightening than a highly questionable pragmatics-based distinction between mathematical and factual concepts.

BETH at first found the interpretation of intension in formalized theories extremely problematic: "it would be very difficult to apply the distinction [between extension and intension] in the domain of pure logic and mathematics" ((1966), p. 466)[19]. In a later attempt (1962) to explicate extension and intension within the framework of model theory, his formulation of intension is a referential one (relative to $\varrho_{\mathcal{M}}$) and satisfies neither (D 1) nor (D 2). The same is true of MONTAGUE's recent proposals (1968). MARTIN (1964) defines intension in terms of a referentially explicated notion of analyticity. SUSZKO (1967) gives half a dozen or more formal explicata for intension and extension, but does not point out the particular combination (that given in Section 1.3) which seems to us so natural, fails to emphasize any syntactical aspect of intension, and even presents as a "result" a distorted version of (D2), namely that if $\text{Int}(F) \subseteq \text{Int}(G)$ then $\text{Ext}(F) \subseteq \text{Ext}(G)$ (ibid., p. 21).

Insofar as the Port Royal formulation of the notion of intension may be connected with that branch of medieval logic designated in Section 1.1 as the theory of consequence, while that of extension falls within the theory of supposition, we feel justified in claiming, for the

---

19 BETH goes on to say that "in pure logic and mathematics no reasonable criterion for the identity of intensions is available, except the identity of the corresponding extensions, and that is exactly the criterion which is incorporated into the Comprehension Axiom" (ibid., p. 466). BETH's mistake here is to consider the extension of a set S (more properly, of a set letter S) in a formalized theory to be the "set of values" of the variable x which satisfy the formula $x \varepsilon S$, without making explicit reference to some one model of the formalized theory, in the universe of which x is to take its "values". When the Axiom of Comprehension is adopted in a formalized set theory, it expresses a state of affairs which must hold in every model of the theory: this says much more than the same assertion made relative to one model, which is presumably the case when discussing the "set of values" of a syntactical variable.

The Axiom of Comprehension imposes a syntactically expressed con- condition on all possible models of the set theory, and as such is in fact, in our view, a properly intensional condition. BETH's early unsophisticated handling of the model-theoretic apparatus in discussing extension and intension in terms of values of variables is also evident in his (1956), pp. 42 ff. The apparent dependence of identity of intensions on identity of extensions disappears once one notices that model theory can serve as framework for a proper explication of extension and intension as well as for a referential one, as we have shown.

notion of intension, an autonomy from referential considerations as
great and of the same nature as the autonomy of proof theory relative
to model theory; and in maintaining such an autonomy as an important desideratum for the explication of these notions.

We consider, then, that referential explications of intension trivi-
alize a potentially more interesting doctrine, which may be taken to
straddle both proof theory and model theory. Autonomy of intensions
relative to reference is especially desirable if the doctrine of intension
and extension is to be applied to natural languages or factual
theories (e. g., those of physics), where (linguistic) convention and
(referential) fact both hold leading roles. A syntactical acceptation of
intension also offers richer perspectives, for example, in the case of
semantically inadequate languages.

One could go so far as to claim that the essential business of any
logic is to elucidate intension and extension, that is, to explicate the
relations of entailment and reference and their interconnections. As
far as providing a *formal* clarification of reference, model theory has
proven entirely satisfactory relative to what is usually taken to be the
classical notion of entailment. But such a high degree of articulation
as that obtained in Section 1.5 is not necessary to an adequate explication of extension and intension, and in some cases not even
justified, as illustrated in 1.8; extension and intension can be meaning-
fully discussed in any language in which independent relations of
entailment and reference can be recognized, and validated together
through (D1) or (D2).

This independence of intensions relative to referential considerations, and, conversely, that of extensions relative to entailment —
despite (D2), it is not intensions, but reference which determines
extensions — will be fundamental in our discussion below of the
notion of meaning. We will indeed propose that meaning and truth
alike are best construed as two-dimensional, as possessing both a
referential and a linguistic component, neither of which is reducible
to the other.

2. Meaning

2.1 Correspondence and Coherence Views

For a start on the subject of meaning, it would be difficult to find a more open and frank presentation than that of LEWIS (1946). Therein he follows PEIRCE in holding that "the essentials of the meaning situation are found wherever there is anything which, for some mind, stands as sign of something else ... The genuine signification of meaning is that in which A means B if A operates as representing or standing for B or as calling B to mind" (ibid., p. 72). For our subsequent discussion, an explication of the notion of meaning will consist of an interpretation of LEWIS' description. We will only discuss meaning in relation to linguistic or conceptual things (constructs), designated within some language L[20].

Any explication of meaning which, for a construct F in L, proposes entities outside of L, i. e., extra-linguistic entities, as that which F "represents", or "calls to mind", we shall call a *correspondence* explication of meaning. Any interpretation of meaning which does not employ extra-linguistic entities in this way will be called a *coherence* explication. Of course, coherence theories of meaning may employ extra-linguistic methods or operations to indicate, or point towards, meanings; a pure coherence approach to meaning, relying on no extra-linguistic apparatus whatsoever, would be obviously impossible, inasmuch as meanings are always understood relative to some background theory or language.

---

20 It is convenient now, and in the following chapters, to broaden our usage of symbols such as L, F, etc., beyond their ranges in Chapter 1, to designate informal languages, constructs, and so on.

Clearly, both the coherence and correspondence views fit PEIRCE's picture. Among various correspondence theories of meaning, we will deal mainly with *referential* theories, which we understand to present existing (physical or Platonist) objects as the things signified or meant: other correspondence views, such as ideational theories, which offer "ideas" as the end of signification, and behaviourist theories, offering a stimulus-response view of meaning, have not recently been seriously applied to mathematics. Even in the rarefied atmosphere of foundations research and mathematical logic, it is a vaguely referential, Platonist view of meaning which is most commonly advanced, to such an extent that we may call it the received view of meaning in mathematics.

This referential view owes its vitality to two main factors: the success of the "reduction" of mathematics to sets completed by the end of the last century, and the advent of Tarskian model theory as a formal semantics for formalized mathematics. Meaning in mathematics has become bound to sets to the point that, to the dismay of some, sets have become the vehicle of mathematical communication right down to the level of kindergarten. Of course, there are good reasons for this: finite sets are simple to conceive, have physical referents with which set-theoretic operations can be physically mimed, and so on. However, it is this very connection (through finite sets) of set theory with the real world, together with the above-mentioned successful "reductions" and "semantics", that encourages the construal of sets as providing a Platonist universe for a hazy referential view of meaning in mathematics.

The highly prevalent attitude in mathematics, particularly in contemporary logic texts, is that one specifies a "meaning" for the formulas of a formalized theory, by presenting a set-theoretic model of the theory. This is illustrated by TARSKI himself: "A non-categorical set of sentences (especially if it is used as an axiom system of a deductive theory) does not give the impression of a closed and organic unity and does not seem to determine precisely the meaning of the concepts contained in it" (1956, p. 311). But under such a narrowly referential view of meaning in formalized theories, we must deny the entirely satisfactory axiomatizations of the theory of groups any "organic" unity, and deny precise meaning to the PEANO axioms or to any other formalized theory with a countably infinite model. This is clearly absurd.

The assumption that the study of the semantic aspect of formalized theories — taken in its narrow, referential sense as the study of the relations between these theories and some "reality" — can properly be carried out within the language of set theory, withers upon closer scrutiny. We have seen in Section 1.3 that the notion of reference can be explicated in several distinct ways in model theory depending on whether one adopts the truth-set or the Tarskian explication of object, and on whether one adopts ϱ_M or $\varrho_{\mathcal{M}}$ as explicatum of reference (each is appropriate in different contexts, ϱ_M for theories where one intended model stands out, $\varrho_{\mathcal{M}}$ for theories which aim at generality); but most important of all such considerations, there is not one model theory but several distinct model theories, according to the underlying set theory in which model theory is carried out.

This essential relativity of model theory undermines its validity as a vehicle for the explication of reference in mathematics. The fundamental diversity of set theories, recognition of which is forced upon us by the recent independence results in set theory (cf. COHEN, 1966)[21], invalidates also the premise necessary to the referential view, namely the attribution of some kind of ontological quality to the theory of sets. But let us postpone arguing the de-ontologization of set theory in detail until the next chapter.

We conclude that the received view of meaning for formalized theories is superficial and inadequate. Set theory does furnish a convenient conceptual scheme for the formal expression of basic logico-semantical relations, but fails to provide a possible objective pole for a referential view of mathematics, as standing for mathematical "reality". Model theory can be only colourfully described as "a study of the interrelations of a language — usually a precise logical symbolism — and the 'reality' this language represents" (HINTIKKA, 1969, p. 2): it cannot be considered to constitute the semantics of mathematics, in the narrow, referential sense of "semantics".

Failure to keep the set-theoretic "semantics" of formalized theories in proper perspective results in the wildest philosophical declarations, ranging from the well-known extravagances based on the LÖWENHEIM-SKOLEM Theorem[22], to interpreting GÖDEL's Completeness

---

21 Witness the title of an abstract of SOLOVAY: "$2^{\aleph}$ Can Be Anything It Ought to Be", in ADDISON, HENKIN, and TARSKI (1965), p. 345.

22 That the SKOLEM hoax is at last being caught up with by philosophers is evidenced by HART (1970).

Theorem as "a modest but precise expression of the rather sweeping
claim that mathematical existence is nothing more than consistency"
(LYNDON, 1966, p. 55), or, more sweepingly still, as saying "essentially, that we cannot talk consistently without talking about something" (HATCHER, 1968, p. 39). A more critical examination of the
philosophical ground for such metaphors would immediately reveal
their doubtfulness.

If the referential view of meaning for formalized theories is to
such a point questionable, what can be hoped for from the referential view when applied to non-formalized mathematical or other
discourse? Yet DAVIDSON (1967) has recently again proposed that
the general theory of meaning must be reduced to the theory of
truth, which he construes referentially. As he begins with the assumption that the notions of meaning and of semantical interpretation are identical, it is not surprising that he concludes that "the
obvious connection between a definition of truth of the kind TARSKI
has shown how to construct, and the concept of meaning [is that]
to give truth conditions is a way of giving the meaning of a sentence" (ibid., p. 310)[23].

But while DAVIDSON's stated ambition is to provide a general
theory of meaning which will cover, in particular, natural languages,
he admits that a satisfactory definition of truth (TARSKI's) has been
obtained only for formal languages, and that prospects for a semantical theory of natural languages are poor (ibid., p. 310). To this
we would add that even for formal languages, TARSKI's explication
of truth remains formal, and a properly referential, extra-linguistic
theory of truth remains to be exhibited: indeed, in Section 2.6 we

---

23 We find DAVIDSON's "extensionalization" of the problem of
determining meaning (ibid., p. 309) impossible to follow. That DAVIDSON
construes truth referentially is explicit: "Indeed since a TARSKI-type truth
definition supplies all we have asked so far of a theory of meaning, it is
clear that such a theory falls comfortably within what QUINE terms the
'theory of reference' as distinguished from ... the 'theory of meaning' ...
A theory of meaning ... is an empirical theory, and its ambition is to
account for the workings of a natural language" (ibid., pp. 310—311).
Incidentally, this last sentence invalidates a report by HINTIKKA of
DAVIDSON's article, where he presents DAVIDSON as arguing that "in
first-order languages the theory of truth and reference is (or can be put
to use as) the theory of meaning" (HINTIKKA, 1968, p. 15). DAVIDSON in
fact spends no time at all on first-order languages.

will argue against the pure correspondence view of truth in mathematics generally.

Even to most *habitués* of the Vienna Circle, it soon became clear that, even for factual theories, where one has an arsenal of physical objects at one's disposal, one should know better than to rest meaning entirely on reference: "the cognitive meaning of a statement in an empiricist language is reflected in the totality of its logical relationships to all other statements in that language and not to the observation statements alone" (HEMPEL, 1950, p. 181).

But the pendulum can swing too far the other way, towards a pure coherence view of meaning, as advocated, for example, by GOBLE: "the meaning of an expression in a language is constituted by relations which obtain between it and other expressions of the language ... the relations relevant to meanings are consequence relations ... the meaning of an expression is determined by its 'inferential role'" (1967).

It is clear that the coherence view cannot account adequately for the meanings of constructs in a factual theory. BUNGE's observation, that the working definition of "mammalia" currently accepted by zoologists exclusively involves osteological properties related to the lower-jaw-middle-ear complex (1967, vol. 1, p. 67), amusingly illustrates the irreducible role of reference in the determination of meaning in factual theories[24]. As for the problem of comparing meanings, formal analogy is the only tool offered by a pure coherence view for the all-important activity of inter-theoretic explanation in factual science: this is unacceptable (cf. beginning of Section 4.1).

As a satisfactory explication of meaning should deal adequately with meaning in factual theories (which refer to facts) as well as in non-factual theories (which refer to nothing real), we must therefore reject this Wittgensteinian coherence view of meaning. But even for non-factual theories, the limitative theorems on formalization (e. g., GÖDEL's incompleteness theorems) can be interpreted to the disadvantage of the pure coherence view, insofar as they inhibit the crystallization in formal languages of the relations of entailment presumed to hold in the non-formalized theories.

---

24 Such examples defy QUINE's catch-phrase: "Meaning is what essence becomes when it is divorced from the object of reference and wedded to the word" (1953, p. 22). Some divorce! Some wedding!

2.2 Meaning as Intension/Extension

We have argued above that for many theories, meaning is not determined solely through the consideration of relations of reference, nor solely through relations of entailment, but through a combination of both, concluding that neither a pure referential, correspondence view nor a narrow coherence view of meaning is satisfactory. It cannot be denied that even in everyday discourse, meanings are determined partly by convention, partly through ostension.

The particular combination of reference and entailment appropriate to a given theory will depend more heavily on reference according as the theory deals more directly with facts, and will depend more on relations of entailment according as the theory is highly sophisticated and assumes a number of other theories which mediate between it and the facts. On the other hand, one can easily conceive of primitive theories or languages where meanings are determined through reference alone, while emerging theories in mathematics may lack any mediating theories, or are so far removed from factual interpretation in their inception and initial development, that their meanings may be considered as determined exclusively through relations of entailment.

A dualistic explication of meaning, offering both a correspondence component and a coherence component, appears inevitable if one is to encompass in a single explication such diverse meaning-situations. This dualist view is already suggested in HEMPEL (1950) (cf. the quotation above). It is also explicit in LEWIS (1944) and (1946), where he develops PEIRCE's picture in several different ways (we will formally explicate most of these in Section 2.5). BUNGE (1967) presents the view that the meaning of a construct must be explicated by the pair consisting of its intension and its extension.

BUNGE's proposal expresses a shift in emphasis away from the diversity of the modes of meaning, towards the recovery of a unified conception of meaning: a unity in diversity, but not in an unstructured diversity, rather a diversity in which the component parts (extension and intension) exhibit a certain degree of interrelation. Under this dualistic view, the problem of explicating meaning bifurcates into the distinct problems of explicating entailment and reference. But this bifurcation is not absolute: in the process of explicating entailment it is necessary to keep an eye on reference, and *vice-versa,* in order to ensure a sufficient degree of unity to the whole.

Recognition of independent but suitably articulated linguistic and referential components of meaning is, of course, as old as the doctrine of intension and extension itself, which can be described, as in Section 1.1, as a move to unify the medieval theories of consequence and supposition. What *is* surprising, instead, are the continued attempts to explicate meaning solely through a linguistic (or coherence) framework, or solely through a referential (or correspondence) theory. It seems obvious that a dualistic explication of meaning holds out greater promise.

As an improved approximation, then, we will consider the *meaning* of a construct F, relative to a certain language or theory T, to be given by the set of constructs entailed by F in T, together with the set of objects referred to by F. That is, we explicate the coherence, or linguistic, component of meaning by intension, and the correspondence, or referential, component by extension. The determination of meanings will then depend on the ability to recognize relations of entailment in T, and to apply the relations of reference congenial to T. As we underlined earlier, extensions and intensions need not exhibit a degree of articulation as high as that of classical model theory, as described in Section 1.5.

A more recent proposal of BUNGE departs from this limpid scheme, by superimposing on this dualistic framework the view that the meaning of a construct F has to do with *all* the constructs which are logically related to F. BUNGE (1972) defines the *sense* of a construct F to be the union of the antecedents of F and of the consequents of F, that is, the sense of F is the set consisting of those constructs G such that either G entails F or F entails G. Before seeking to apply our view of meaning to mathematics, let us examine this further possible move.

It is natural to call the set of all logical antecedents of a construct F the *co-intension* of F. The sense of F is then the union of its intension and its co-intension. BUNGE's motivation for including the co-intension of F in the sense of F seems to lie in his subsequent introduction of the notion of the *gist* of F, the latter being a certain restricted subset of antecedents of F which constitues the "essential" sense of F.

We feel that the notion of a core intension of a construct (cf. Section 1.7) makes this use of co-intensions unnecessary. Since the conjunction of the constructs in a core intension of F is logically equivalent to F, such a core intension could be taken to be the gist of F. This would be advantageous not only for reasons of economy, but also because a construct in a core intension of F need not be an antecedent of F, i. e., can be logically strictly weaker than F, which is not the case with the constructs in BUNGE's gists. This seems to us to be a sound desideratum to adopt if the constructs specified in the gist or core intension of F are to be in some sense weaker, or more easily understandable, than F. For example, {"rational", "animal"} is a core intension of "man". Or again, extending the definition of intension to sets of formulas as in Section 1.6, one core intension of a finitely axiomatized theory is the set of its axioms, another would be any different, but logically equivalent, finite set of axioms. The notion of core intension renders, in this way, that of co-intension dispensable.

Confusion can result from an overexuberance of logical *genres.* If it is considered important to bestow some algebraic order or orientation upon intensions, or "senses", then it is preferable not to consider a construct to point two ways at once (to antecedents as well as to consequents) in a single logical dimension. Most of the algebraic properties and interrelationships which hold for extension and intension, as developed in Sections 1.3 to 1.5, do not hold for BUNGE's notion of sense. Because the chains of antecedents of a construct proceed in a direction opposite to that of the chains of its consequents, even a relation of set-theoretic inclusion between two distinct senses would be unusual[25].

We prefer to consider the linguistic sense of a construct F to be that (those constructs) to which F linguistically *points towards,* and its referential sense to be that (those objects) to which F points towards referentially, or *refers.* Serviceable calculi of intensions and extensions, and a satisfactory degree of articulation between them, can only be obtained at the price of restricting these pointers each to a single direction.

---

25 We might note here that under our explication of meaning, the meaning of a construct G is rarely contained in that of another construct F. For if such is the case, then G can only differ from F linguistically, in its intension, since from $\mathrm{Ext}(G) \subseteq \mathrm{Ext}(F)$ and $\mathrm{Int}(G) \subseteq \mathrm{Int}(F)$ it follows, by (D 2), that $\mathrm{Ext}(G) = \mathrm{Ext}(F)$.

2.3 Meaning of Constructs in Mathematical Theories

It is necessary here to adopt a more careful vocabulary for the discussion of the process of formalization. We distinguish between a mathematical theory in the rough, such as, say, Euclidean geometry as developed in the *Elements*, or as expounded in a classroom, or as conceived by an active geometer, and a mathematical theory in the formal sense; we call the latter a formal theory, as usual, and the former a *natural* theory[26]. When a formal theory T is judged to render faithfully a natural theory T′, we will call T a *formalization* of T′. Our previous practice of referring to "formalized" theories conflated T′ and T in such a situation.

The discussion of meaning in a given language or theory must take place in a metalanguage or background theory. The background theory can vary widely according to the object theory: for the syntax of formal languages, a sparse background theory consisting essentially of classical arithmetic is appropriate; for formal semantics, a set theory is used; and so on. For natural mathematical theories, the background theory consists generally of a fragment of a natural language together with rather rigorous rules according to which inferences are made. Hopefully, the background theory provides a basis for the common understanding vital to the communication, development, and application of the object theory.

In absolute fact, things do not run so smoothly. There is a regression of background theories, eventually fading into areas which are the proper concern of epistemology and psychology. But we need not pursue so far, as there is evident in mathematics a persistent phenomenon of objectivity, in that it has always been possible for mathematicians to come to agreement, or at least to clearly elucidated disagreement, over what relations of entailment are valid in a given natural theory[27].

---

26 This terminology must not be taken to indicate a premature *parti-pris* for a realist view of mathematics. It is rather meant to parallel the accepted terminology for languages.

27 Irreducible disagreement over the validity of a certain relation of entailment is rare, and in any case, rather than persist as an insurmountable crisis, simply divides the theory into two distinct natural theories, each with its own rules of inference; we have classical and intuitionistic versions of mathematical analysis, for example. In the sequel,

This phenomenon has been variously described as a *community of understanding* (MYHILL, 1951), as a *mathematische Tatsächlichkeit* (BERNAYS, 1950), or again as *informal rigour* (KREISEL, 1967a), and has served as a basis for the rejection of the non-standard countable models of set theory (MYHILL), as source of inspiration for mathematical foundations (KREISEL), and even as ground for discussing existence in mathematics (BERNAYS). We will take up these themes in the following chapter, but for our present purpose the following related observation is essential: whenever agreement over what relations of entailment hold in a natural mathematical theory has proven particularly recondite, it is ultimately through recourse to formalization that a consensus is re-established. Insofar as they are clear, the relations of entailment in a natural theory are formalizable, though not necessarily in HILBERT's restrictive sense.

We view formalization, then, as the universally accepted means through which the meaning of the constructs in a natural mathematical theory can be explicated. Though formalization explicates relations of entailment only, we find this acceptable, since, save perhaps at the psychological level, there are no mathematical objects. Meaning in natural mathematical theories has, therefore, no referential component, and reduces to its linguistic, or intensional (or, if one prefers, conceptual) component. It is not on the basis of some ontology of sets that SKOLEM's paradox (the existence of countable models for set theory) is explained, but through the clear understanding of the implications of CANTOR's Theorem as proven in a formal first-order set theory, and as interpreted in a set-theoretic model (cf. LYNDON, 1966, p. 62). Even GÖDEL, whose Platonist leanings are well-known, considers the *explanation of the content of the theorems of a natural theory* to be the proper goal of formalization (cf. MEHLBERG, 1960, p. 397) [28].

---

then, when we use "relations of entailment", in the plural, we mean various instances of one and the same relation of entailment, much as one conveniently uses "relations of order" to designate various instances of a single order relation, such as "$2 < 3$" and "$1 < 4$".

28 GÖDEL in fact is reported by MEHLBERG as comparing formalization in mathematics with the formalization of physical theories in this respect (ibid., p. 397), a comparison which we find perilous insofar as relations of reference are of importance in physics.

Above all, we steer clear of rhapsodizing over formalization as a bridging together of thought and "object", in the style of, say, LADRIÈRE:

We are not contending that natural mathematical theories are, before formalization, devoid of meaning, but that since the tool universally agreed upon by mathematicians for making meaning explicit is formalization, it is appropriate to explicate meaning in mathematics through its good services. One could simply explicate the meaning of a construct F in a natural mathematical theory as the intension of F relative to the natural entailment relation in that theory; but this would destroy the possibility of a consensus, as what one is permitted to define and deduce in a natural theory, for example in a predicativist, constructivist, or finitist mathematics, is often unclear before formalization. Even intuitionists have been led to formalize: it is the formalization of intuitionist mathematics (and intuitionist logic) which has led, after several decades of neglect, to recognition of the intuitionist view as worthy of interest by mathematicians at large. Formalization is a common ground upon which mathematicians of all philosophical tendencies can meet to agree or disagree — rationally.

If the formal theory T is recognized, in some background theory, as a formalization of the natural theory T', then there must, in an informal but precise way, be established in the background theory a correspondence between the constructs of T' and those of T, and a correspondence between the relations of entailment in T' and those in T, such that those relations of entailment which were judged valid in T' correspond, upon substitution of the appropriate constructs, to relations of entailment in T which hold in the sense defined for formal theories in Section 1.3. In brief, T must be in some non-formalized but precise sense a model, or translation, or interpretation of T'. We will refer to such an informal correspondence between natural and formal theories, or to a similar correspondence between two natural mathematical theories, as a *quasi-translation*.

It will be convenient in the sequel to also use this term to designate any of the more or less precise translations, comparisons, analogies, and other informal and partial meaning-preserving correspondences, by means of which we come to understand, develop,

---

"The project of complete formalization is ... to constitute the total mathematical object ... The distance separating, in research, thought from its object, collapses; mathematical being is then simply and immediately present ... The total [formal] system is not pure objectivity, but a synthesis of objectivity and subjectivity" (1957, pp. 408—410).

and apply a theory. The notion of quasi-translation thus has one foot in pragmatics; different users of a theory or language may employ different quasi-translations of the theory to arrive at a like end. Let us call the totality of all the quasi-translations of a theory the *heuristic component* of that theory.

The heuristic component of a natural theory generally remains somewhat indefinite and fluctuating, even as does its background theory, which, through the interpretation it gives of the natural object theory, is itself part of the object theory's heuristic component. Because of its vague and variable nature, it would be improper — and lead to an endless regress — to utilize the heuristic component directly in explicating meaning in a natural theory. But within our framework the heuristic component of the natural theory T′ is indirectly taken into account, in the judging of the adequacy of the formal theory T as a formalization of T′.

The relations of entailment in a natural mathematical theory are sharp, or must be sharpened through formalization; but the background theory, and the heuristic component generally, may well be hazy in other respects. The various motivations for positing entailment relations can be more or less obscure and even contradictory, but the presumed relations of entailment are subsequently objectively accredited or not. The vagueness surrounding even such an apparently simple mathematical concept as that of a polyhedron can nonetheless lead, as LAKATOS (1963) has cleverly shown, to a surprising chain of improved analyses, though the crises encountered in the concept's evolution are never due to faulty proof steps. The relations of entailment — the intensional component of mathematical meaning — are what crystallizes out of a natural mathematical theory under the action of diverse and possibly quite inexplicit heuristic strategies. We will develop further the notion of heuristic component in the following chapters.

In the process of formalization of T′ by T, the particular kind of quasi-translation under inspection is of a relatively precise nature, in that it must preserve, as far as they can be definitely discerned, the relations of entailment which hold in T′. The notion of model current in mathematics in the last century was a quasi-translation of similar precise sort: a correspondence was established between the primitive constructs of a natural theory T″ and the constructs of another natural theory T′, in such a way that the axioms of T″ were "translated" into theorems of T′, e. g., the various models of non-Euclidean geometries in Euclidean geometry, or the modelling of the complex numbers in the reals.

This type of inter-theoretic relation was already explicated for formal theories by CARNAP (1939): given formal theories T'' and T', both expressed in the same language L, we will call any mapping from the set of atomic formulas of T'' into the set of formulas of T', such that the induced mapping on the set of all formulas of T'' maps axioms of T'' onto theorems of T', a *syntactical interpretation* of T'' in T'. In contrast, to follow common logical parlance, we will henceforth refer to a Tarskian set-theoretic model of T'', as used in Section 1.3, as furnishing a *semantical interpretation* of T''.

But more is to be required of a formalization T of T' than that it result from an informal syntactical interpretation. This particular brand of meaning-preserving correspondence should also be conservative, in the sense that no "new" relations of entailment should hold in T among images of constructs from T'; that is, if F and G are the images in T of constructs F' and G', then we should not have $\vdash_T (F \to G)$ without having that F' entails G' in T'. The notion of syntactical model, or interpretation, does not include this conservative condition, because it was not developed to explicate the notion of inter-theoretic meaning-preserving correspondence, but rather for use in structuring relative consistency proofs, for which preservation of theoremhood is of import in one direction only.

Let T'' and T' be formal theories expressed in the same language L. A *faithful translation* τ of T'' in T' is a syntactical interpretation of T'' in T' such that, if the formula F'' of T'' is mapped by τ onto the formula F' of T', then

$$\vdash_{T'} F' \text{ (if and) only if } \vdash_{T''} F''.$$

It follows that, under a faithful translation, intensions of formulas (as defined in Section 1.3) are rigorously preserved, in the precise sense that for any two formulas F'' and G'' in T'',

$$\vdash_{T'} [\tau (F'') \to \tau (G'')] \text{ if and only if } \vdash_{T''} (F'' \to G''),$$

since by definition of τ, $\tau (F'' \to G'')$ is $\tau (F'') \to \tau (G'')$.

A relatively objective explication of the meaning of mathematical constructs can now be obtained by defining the *meaning* of a construct F' in a natural theory T' to be the intension, as explicated in Section 1.3, of the corresponding construct F in some formalization

T of **T′**. The meaning of $F′$ is thus relative to the particular formalization **T**.

It is essential that we be able to compare meanings taken relative to one formalization, say τ_1: **T′** → **T₁**, with meanings taken relative to another formalization τ_2: **T′** → **T₂**, in order to be able to decide, in some precise way, whether meanings remain essentially the same, or are changed, under the different formalizations. But this should present no difficulty, in that if τ_1 and τ_2 are both valid (informal) faithful translations, then they should induce a partial formal faithful translation τ: **T₁** → **T₂** which is an isomorphism between the restrictions of the intensional calculi of **T₁** and **T₂** to images under τ_1 and τ_2 of the formulas of **T′**, i. e., between the intensional subcalculi of **T₁** and **T₂** constituted by the images of formulas in **T′**. Schematically, the diagram below should commute

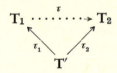

on meanings (τ_1 followed by τ should agree with τ_2 on intensions).

The process of formalization can now be explicated as the establishment, in a certain background theory, of an informal faithful translation of a natural theory **T′** in a formal theory **T**. For a successful formalization consists precisely in obtaining a correspondence between the constructs of the natural theory and the formulas of the formal theory such that syntax (how constructs are built up out of other constructs) and meanings (intensions, relations of entailment) are, as far as possible, rigorously respected. That there are, for virtually every natural theory in mathematics, formalizations which are universally accepted by mathematicians, is perhaps the most conclusive argument in favour of construing meaning in mathematics as devoid of any extensional, or referential, component.

2.4 Meaning in Formal Theories

We may now explicate meaning for constructs in a formal mathematical theory **T**, by taking the meaning of a formula F in **T** to be Int (F). There is no extensional component to meaning in formalized theories, because meanings in formal theories must be derived from the natural theories which they formalize, and there is no objective referential component of meaning in natural mathematical theories.

Our explication of meaning in Section 2.2, as the pair intension/extension, was a general one: when the problem of explicating meaning is particularized to a certain restricted domain, then one of its dual components may be inactive. Situations can equally well be imagined where the intensional component is inactive, for example, in a very primitive language, where meanings are entirely indicated through ostension. Here, the intensional component cannot properly be filled by anything more than intensions which are wholly dependent on reference, i. e. by comprehensions (recall Section 1.4), and in such a case is no more than a passive mirror-image of the relations of reference.

The popular practice of describing meaning in formal theories exclusively in terms of extensions of formulas, defined in Tarskian models, is due in part to the failure to recognize the proof-theoretic explication of intension in formal theories: "The delicate point in the formalistic position is to explain how the non-intuitionistic classical mathematics is significant, after having initially agreed with the intuitionists that its theorems lack a real meaning in terms of which they are true" (KLEENE, 1952, p. 57). Without being narrowly formalistic, we feel that formal (classical) mathematics can be profitably construed as deriving its significance from the natural theories, from the various heuristic *milieux* of mathematics, from which it arises through the process of formalization as described above. This significance rarely — perhaps in some parts of what is nowadays called "finite" mathematics — has any referential component, so certainly cannot be described in terms of "real" meaning: in our opinion, not even intuitionistic mathematics has succeeded in doing *that*.

Formal theories are simply not approached as arbitrary, rule-directed games, played in a meaning vacuum, which vacuum must be filled by an ingeniously contrived semantics. A particular move in these games is recognized as a proof, and carries some substantial significance: this is attested by the care taken to provide soundness, or validity, theorems for any proposed formal semantics.

The importance ascribed to referential explications of meaning in formal theories may also be due to the limitative results (incompleteness theorems, LÖWENHEIM-SKOLEM theorems) concerning formalizations, especially first-order ones. These limitative theorems arise in the study of the formal semantics of a formalization, that is, in a background theory usually containing a theory of sets, and which is strictly stronger than the background theory commonly considered as sufficient for the formal first-order explication of entailment relations in a natural theory. It is not surprising, then, that in the theory of models distinctions can be made which escaped syntactic considerations. If a formalization is to be critically examined from the more powerful perspective provided by the theory of models, it becomes equitable to utilize set-theoretic semantics for a more detailed explication of meaning in such a broadened context.

Suppose, then, that T is a formalization of T', and that N is a model of T (in a particular set theory) which is judged to possess properties which, in the light of the explication of meanings already accomplished through the formalization, reflect more (or most) precisely the meanings ascribed to the constructs of T': then N may be singled out as the *intended* model of T, and the explications of meaning provided by the relations of entailment in T may be seconded by the relations of "reference" relative to ϱ_N. This procedure would be appropriate, for example, in a first-order formalization of arithmetic. On the other hand, in a formalization of group theory, no particular model seems relevant in specifying meaning beyond the intensional relations already available in T', and $\varrho_\mathcal{M}$ would be the only appropriate "reference" relation here, since extensions taken relative to $\varrho_\mathcal{M}$ add no insight into meaning which is not already reflected in intensions relative to η.

An intermediate situation can arise, say in a higher-order formalization of set theory, where it is natural to distinguish a set $\mathcal{N}$ of *standard* models (HENKIN, 1950): the appropriate "reference" relation here would be $\varrho_\mathcal{N}$. Of course, if T is categorical, then, as in the case of the formal theory of groups, nothing new is to be gained by tacking a referential component onto meanings, as the calculus of extensions taken relative to any model is in such a case only a passive mirror-image of (dual isomorphic to) the calculus of intensions [29].

---

29 These descriptions contrast with the usual version, according to what we called the received view, that to specify a model for T is to bestow a meaning on the formulas of T: for us, to specify a model is to *complement* meanings which are already (partially) recognized.

Thus, in this broader context, the entire apparatus of extension and intension as explicated in Chapter 1 can be brought to bear on the problem of explicating meaning in T'. This explication of meaning will be doubly relative, firstly to the formalization T, secondly relative to the particular model theory chosen for T. The ontological overtones of set theory, and the fact that ontological preoccupations springing from the heuristic component of T' may be involved in distinguishing between standard and non-standard or between intended and unintended models, might support describing the specification of an explicatum for "reference" as the provision of an "ontology" for the theories in question.

But this common practice must be viewed with circumspection, in that a semantical interpretation is nothing more than a quasi-translation between languages, a formal language of syntax and a non-formalized language of sets, and only possesses the characteristics, the appearance, of reference. If the complementary properties which make so valuable the specification of an intended model as an "ontology" for the formalization T were considered in themselves and not only relative to T, then it would be proper to clarify this fact by specifying a formalization of T' directly in a formalization S of the set theory in which the model theory is being construed. The meanings of the constructs of T', taken relative to S, would then stand stripped of their supposed "referential" components, the valuable properties of the latter being presumably expressible in the intensional component taken relative to S. The artificiality of the "ontology" provided by S, as being an ontology only relative to the weaker T, can in this way be recognized.

Our position thus leads us to regard TARSKI's development of model theory not only as an advance in the formal study of reference, but as an advance in the explication of meaning as well. Indeed, the dualistic view of meaning which we have thus far developed opposes QUINE's separation of meaning and reference (cf. 1953, p. 130 and p. 21). We feel that both entailment and reference have essential roles to play in the determination of meaning in general, roles which can even be recognized (though only formally) in our explication of meaning in mathematical theories. We hope that our discussion above, and further arguments to follow, will show up QUINE's dichotomy as mistaken and sterile.

2.5 C. I. Lewis on Meaning

In his review of LEWIS (1946), HEMPEL finds that "an adequate appraisal of the value and even of the possibility of the analytic approach advocated by him requires the construction of a formalized theory in accordance with it" (1948, p. 41). We will partially fulfil HEMPEL's request by explicating many of LEWIS' concepts in the framework developed in Chapter 1.

LEWIS' programme in Book 1 (ibid.) is to explicate the traditional notion of intension, in order to justify the traditional conception of analytic truth as truth determined by meanings alone. In isolating that sense of intension which can provide sufficient guarantee for this notion of analyticity, LEWIS is led to distinguish between various modes of meaning for terms, a *term* being any linguistic expression which names, or applies to, a thing or things of some kind: "The denotation [or *extension*] of a term is the class of all actual things to which the term applies. The *comprehension* of a term is the classification of all possible or consistently thinkable things to which the term would be correctly applicable. The *signification* of a term is that property in things the presence of which indicates that the term correctly applies and the absence of which indicates that it does not apply ... the *intension* [or *connotation*] of a term is ... the conjunction of all other terms each of which must be applicable to anything to which the given term would be applicable" (ibid., p. 39).

LEWIS insists on the "must" in his definition of intension, emphasizing that for a term G to be part of the intension of a term F, we must have that the applicability of G to a thing "entails" or "strictly implies" the applicability of F to that thing (ibid., p. 40) [30]. While entailment, then, is the relation regulating what is to be of the intension of a term, and while reference relations determine a term's extension and comprehension, LEWIS invokes a third kind of relation — one could call it a *relation of signification* — with which to determine the signification of a term, and says that a term "signifies" its signification or "comprehensive essential character" (p. 41).

---

30 For simplicity, we continue to use our previous notation to designate LEWIS' terms.

Three of LEWIS' modes of meaning find a natural explication in the framework developed in the preceding chapter. Given a language L (or theory T) and the set of models $\mathcal{M}$ of L, we can take the *denotation* of a construct F in L to be $Ext_M(F)$, defined relative to some fixed M in $\mathcal{M}$; the *L-comprehension* of F to be $Ext_{\mathcal{M}}(F)$; and the *intension* of F to be $Int(F)$, all this as in Sections 1.3 and 1.4 [31]. We shall employ "*L-comprehension*" instead of LEWIS' "comprehension" in order to avoid confusion with our different notion of comprehension, set out in Section 1.4. Because of the element of necessity which LEWIS insists upon in his definition of intension, the latter notion is not a comprehension (in our sense) relative to ϱ_M, i. e., relative to what LEWIS calls "extensions". However, LEWIS' intensions are comprehensions relative to $\varrho_{\mathcal{M}}$, for he explicitly assumes that all the languages under his consideration are semantically adequate relative to $\varrho_{\mathcal{M}}$ (p. 47).

The points of discord between our formulations and LEWIS' notions are minor, and, we feel, can even be turned to our advantage. One such departure is that, following our explication, the intension of a construct is a set, rather than a "conjunction". LEWIS, despite his statements to the contrary, appears uncomfortable with the fact that the intension of a term is not an expression, or at least something as close to an expression as possible: he affirms that in any case, a definition "provides brief expression of the connotation of the defined term" (p. 44).

LEWIS' defence of his defining intension as a "conjunction" rests on a doubtful turn of terminology. He asserts that "if a connotation should be thought of as a class, then it might be suggested that a term having zero connotation is one which connotes no other term" (p. 45). HEMPEL has already criticized LEWIS' usage of the phrase "zero connotation" — LEWIS uses "zero" in the sense of "null" in "null class" — and suggests redefining the intension of a term as the conjunction of all those terms which are connoted by it but which are not connoted by every term (1948, p. 41). We find this refinement unpalatable, as it destroys the reflexivity of the relation of connotation (or entailment) and hence also the equi-

---

31 The explication of nonexistent possibles obtained in this way through the notion of *L*-comprehension, seems to us simpler than that proposed by VAN FRAASSEN (1967), while achieving essentially identical results, e. g., universally valid statements are theorems.

valence of desiderata (D 1) and (D 2). We think it preferable instead to remove the source of the misunderstanding, along with LEWIS' sole explicitly stated ground for refusing to construe intensions as sets, by simply replacing the guilty phrase by the more correct terminology "minimal connotation".

Another point of discord between LEWIS' formulations and ours is his usage of the term "classification" in defining the L-comprehension of a term. This terminology had earlier caught CHURCH'S attention (1944, p. 29). He took it as an acknowledgement by LEWIS of the difficulties of conceiving of non-existent but possible entities as members of something like a set, and reproaches LEWIS his passing over such difficulties. HEMPEL also sticks on this point, and suggests that "when LEWIS speaks of 'consistently thinkable things', what has to be consistently thinkable is rather a combination of attributes, or a conjunction of terms signifying them", and hence that "all the concepts in the analysis of meaning for which LEWIS would resort to the concept of [L-]comprehension, might equally well be dealt with in terms of the concept of intension" (1948, p. 40), concluding that CARNAP may have been right in advocating that the concept of L-comprehension might be dispensed with entirely (cf. CARNAP, 1956, p. 67).

We find undesirable such a radical reaction, and also consider it unnecessary, for our explication of LEWIS' L-comprehensions is precisely an extensional, or referential one, in terms of $\varrho_\mathcal{M}$. A referential treatment of nonexistent possibles permits a richer semantical articulation between distinct poles of meaning, as suggested, for example, in the previous section. We will expand on the utilization of $\varrho_\mathcal{M}$ in elucidating the doctrine of nonexistent possibles in Section 2.7.

A last *nuance* which we feel can be turned to our profit lies in the apparent discrepancy between our notion of intension and LEWIS', which arises from comparing our linguistic definition of intension, given in terms of η, with LEWIS' definition, which seems to rely in some essential way on reference. But this dependence on reference is heavily superceded by the "must", "entails", or "strictly implies" phrases with which LEWIS always accompanies his referential definitions of intension, and the truth of the matter is that LEWIS would likely consider our acceptation of intension equivalent to his. For LEWIS explicitly assumes that the converse of (D 2) holds relative to L-comprehensions, i. e., relative to reference as explicated by $\varrho_{\mathscr{M}}$ (ibid., p. 47), hence would agree to construing entailment as induced by reference relative to $\varrho_{\mathscr{M}}$, and so, presumably, to our equivalent linguistic definition of the intension of a term.

LEWIS' preference for a referentially phrased definition of intension for terms can be explained through his choosing to treat both terms and propositions in a like way, that is, as having both extensions and intensions. But since one is inclined to entertain a predominantly referential mode of meaning for terms, and a predominantly linguistic one for propositions, LEWIS adheres to this common usage and refers, in the same breath, to the intension of a term as "the totality of other terms which must be applicable to a thing if the term in question applies", and yet refers to the intension of a statement (any expression which "asserts" a proposition) as "the totality of other statements *deducible* from it" (p. 132, our underlining).

We feel that it would be preferable for LEWIS to intensionalize his definition of intension for terms, just as he extensionalized the notions of extension and *L*-comprehension for propositions in terms of "states of affairs", "possible worlds" and "actual world" (p. 57). We agree that the meaning analysis of terms and of propositions can be carried out in similar ways: but then, recognition of the distinctly linguistic nature of intension in contrast to the referential nature of extension becomes, in our opinion, all the more important[32].

Taking the notion of a model M of a theory T as explicatum for LEWIS' notion of a *possible world* (p. 56), we complete the explication within our framework of three of LEWIS' four modes of meaning, namely the denotation, intension, and *L*-comprehension of terms and propositions. But the signification of constructs, and the attendant notions of property, character, attribute, and state of affairs, escape explicit formulation in our terms.

This is to be expected, as the signification of a construct is intended by LEWIS to aid in discussing the *sense meaning* of the construct as the "*criterion in mind* by which it is determined whether the term in question applies or fails to apply in any particular instance" (p. 43, our underlining), whereas he intends intension to be used in explaining his notion of *linguistic meaning* (p. 132). We can-

---

32 In LEWIS (1951) one finds the same disturbing referential treatment of intension for terms. LEWIS therein attributes to SHEFFER his manner of entertaining propositions as terms.

not posit within our framework a third type of relation, a relation of signification, which relates constructs and "ideas", and which purportedly solves this properly pragmatic problem of how reference is recognized. This problem lies partially within the domain of psychology, and its solution is not yet sufficiently advanced, we feel, to warrant any definitive clarification. On this point we join HEMPEL in balking at the very possibility of a precise formulation of this pragmatic doctrine, in its present primitive epistemological state.

The distinction between sense meaning and linguistic meaning is, nonetheless, the heart of LEWIS' theory: "meanings are not the creatures of language but are antecedent, and the relations of meanings are not determined by our syntactic conventions but are determinative of the significance which our syntactic usages may have" (p. 131).

That part of LEWIS' theory which deals with entailment, reference, and their interrelations is clear and agrees essentially with our explication of meaning in Section 2.2. But the notions of signification and sense meaning, which, along with the attendant ideational apparatus of criteria in mind, properties, states of affairs and the like, are supposed to provide a general setting in which to explain how one solves the subjective, pragmatic problem of determining reference, strike us as ephemeral and insufficiently exact: for "a sense meaning, when precise and explicit, is a *schema;* a rule or prescribed routine or an imagined result of it which will determine applicability of the expression in question" (p. 134, our underlining). This gives us little reason to consider sense meaning, even when precise and explicit, to constitute an objective component of meaning. Further (pragmatic) elucidation of LEWIS' mechanism is mandatory.

The notion of a core intension is the only possible echo in our theory to LEWIS' notion of sense meaning. The setting aside of a finite (or, more generally, recursive) set of axioms as characterizing a certain theory is, in the sense of Section 1.7, to determine a core intension for that theory. The resulting axiomatization acts as an "implicit" definition of the constructs dealt with in the theory. Thus a certain system may be recognized as being a group, in that it satisfies the group axioms. In a sense, then, the set of axioms of an axiomatized theory does behave as a "schema", or "criterion in mind": the same can be said for core intensions of constructs, which can be thought of as explicit "working definitions" for constructs. But such axioms or core intensions become maximally "precise and explicit" under formalization, which is none other than a *linguistic* mode for the explication of meaning[33].

For mathematics, at least, where reference remains strictly formal, it must be agreed that it is preferable to have criteria in language rather than criteria in mind. For example, no dialogue between intuitionistic and classical mathematics was possible until the notions of finitary, constructive, and recursive procedures (rules: *prescribed routines*) were made precise through formalization; then it was discovered, among other things, that classical arithmetic is consistent relative to intuitionistic arithmetic (cf. KLEENE, 1952, pp. 492 ff.), which came as a surprise for some criteria in mind.

LEWIS himself often leaves open the possibility of successfully dealing with all problems concerning deductive systems in terms of linguistic meaning alone, "if logic be straightly enough conceived" (ibid., p. 141; cf. also pp. 163 ff.). We would feel equally optimistic about the possibility of dealing with all problems of meaning in any system, if both logic and reference were straightly enough conceived. But just as whether logic is straightly enough conceived is as much a matter for the logician to decide as it is for the philosopher, so is the problem of straightly conceiving reference a matter to be worked out by the factual scientist, the psychologist, and the philosopher together. And if this is done, the philosopher will find many situations in factual science where precise sense meanings are a myth, due, for instance, to extensional vagueness.

Rather than rest the pragmatic aspect of meaning on the vague ideational notion of criteria in mind, we consider that it would be more direct and more effective to centre such pragmatic concern on the sharpening of criteria in language and of criteria in reference. The objective content of signification is best captured in this way, much as, in mathematics, formalization sharpens and objectifies meanings arising from heuristic components.

---

33 An amusing aside: if the core intensions of formal axiomatized theories are *very* "precise and explicit", that is, if we are dealing with finitely axiomatizable theories, then (and only then) their calculus of extensions, in terms of models, is most highly "exact", in the sense of being perfectly compatible with the set-theoretic operations on sets of models (cf. Section 1.6).

2.6 Truth in Theory and Truth in Practice

Even more familar than the opposition of coherence and correspondence theories of meaning, is the opposition of coherence and correspondence views of truth: a statement can be true because it corresponds to fact, or because it coheres with (is consistent with, logically related to, deducible from) other accepted statements. BURI-
DAN's solution to the liar paradox (cf. MOODY, 1953, ch. 5), that
truth is not a matter of reference alone, but of form (intension) as
well, already weighed heavily against a pure correspondence conception of truth in general. But let us restrict ourselves to the situation in mathematics.

In discussing truth in mathematics, the distinction between theory
and practice is essential. Mathematical activity is often compared
to artistic activity; just as it is important to distinguish between
philosophizing on the end products of art, and philosophizing on the
processes and techniques of artistic creation, so is it important to
distinguish between philosophizing on the theories of mathematics,
and philosophizing on the methods which mathematicians may use
in arriving at their results. Insofar as this distinction can be clearly
made, it offers a peace plan for two opposing factions of philosophers of mathematics.

The dashing in the thirties of the formalist hopes for an absolute
foundation for mathematics supposedly made it impossible to construe mathematical truth as something which could be completely
brought out into the open through the process of formalization. It
is in this light that GÖDEL's incompleteness results are explained to
this day: "Although it is now nearly forty years since GÖDEL showed
that no consistent set of axioms for arithmetic can yield all arithmetical truths as theorems, the notion that in mathematics truth
amounts to nothing more than derivation from axioms seems to die
very hard" (MATES, 1970, p. 303).

As a result of this supposed rupture between truth and proof,
two attitudes became attractive. Either one could take refuge in
a mystical, Platonist correspondence view of mathematical truth,
and construe mathematical theories as issueing from continually
improved intuitive acquaintance with "the facts" — GÖDEL himself
is most commonly cited as a proponent of this school. Or, as in
CARNAP (1937), one could adopt a coherence view, and take truth in
mathematics to coincide with provability. What the correspondence

view gains heuristically, it loses in fuzziness; what the coherence view gains in clarity, it loses in being unable alone to explain mathematical development.

Recently these views have come into open conflict. According to GOODSTEIN, "the number concept is to be found only in the transformation rules of arithmetic and the use of collections as number signs. The question 'what is number' must be replaced by the wider question 'what is arithmetic and what are its applications'" (1968, p. 113); and "By talking of truth in mathematics the empiricist begs the whole question of the nature of mathematics. If truth is divorced from provability then what test of truth remains but an empirical test? ... how can a convention be true or false?" (1969, p. 57). In our last excerpt GOODSTEIN himself is guilty of begging the point for the coherence view: if truth is wedded to provability, then what test of truth remains but truth by proof?

The empiricists GOODSTEIN is referring to are KALMÁR and LAKATOS, both, like POLYA (1962), strongly interested in the processes of mathematical discovery. Fixing his attention on the heuristic aspect of mathematical activity, LAKATOS is led to hold that "informal, quasi-empirical mathematics does not grow through a monotonous increase in the number of indubitably established theorems but through the incessant improvement of guesses by speculation and criticism by the logic of proofs and refutations" (1963, p. 6).

LAKATOS' view can subtly lead to a belief in a transcendent mathematical reality where the yet to be discovered truths of mathematics have their abode. In LAKATOS' sophisticated quasi-empiricism (cf. LAKATOS, 1962), much is made of the fact that in mathematical practice, it is not true truth-values which are "injected at the top", that is, in the axioms, from which truth "flows downwards" into the system, but rather it is false truth-values (counterexamples) which are injected "at the bottom" and which then "flow upwards". But if this is to be more than simply a critique of informal proof, then some source for continually improving hypotheses must be posited, some sense must be given to the dialectic of proofs and refutations: and Platonism provides an alluring answer.

But no one has yet satisfactorily explained how we mortals have access to the Platonist heavens. Upon looking closer, then, LAKATOS' view proves no more enlightening than GOODSTEIN's on the source of mathematical truths. LAKATOS may object to GOODSTEIN's imputing to him "a metaphysical assumption of the existence of a mathematical reality transcending the literature of mathematics, embracing alike the known and the yet to be discovered theorems of mathematics" (GOODSTEIN, 1969, p. 51), but the problem of elaborating a non-coherence (presumably a correspondence) approach to truth in mathematics, with which to deepen LAKATOS' view, would then remain intact.

The LAKATOS-GOODSTEIN debate can be resolved in favour of both opponents (or rather, in their disfavour) if only it is realized that the debaters are talking of different things. GOODSTEIN's conventionalist position arises from his taking the finished product as the object of his philosophy, whereas LAKATOS' quasi-empiricism (and KALMÁR's — cf. KALMÁR, 1967) springs from his taking the goings-on of mathematical research as the heart of mathematics. It is obvious that a coherence view of mathematical truth such as GOODSTEIN's suits philosophizing on the completed mathematical product, and that a correspondence view suits philosophical preoccupation with the processes of mathematical creation. A somewhat similar resolution seems indicated for KREISEL's difference with COHEN, vented in SCOTT (1971).

Thus, whatever its epistemological shortcomings, Platonism at the level of mathematical research is not incompatible with conventionalism at the level of the finished product. Both the correspondence and the coherence views of truth are in their different ways applicable to mathematics. Their spheres of application are not entirely disjoint, in that the partition between end-product and process of fabrication is not perfectly delimitable. It is in these very areas of overlap that the mutation from correspondence to coherence views of truth takes place, or where, in the terminology of Section 2.3, meanings suggested by correspondence views (insights) in the heuristic component of a natural theory crystallize into linguistic definitions and relations of entailment for that theory, that is, into meanings expressible in a coherence framework. The proper sphere for the entertaining of correspondence views of both meaning and truth in mathematics is the heuristic sphere, whereas at the level of theory the coherence view is appropriate.

Among mathematicians, set theorists are notorious Platonists. Philosophers should not begrudge them this luxury — nor even methods of divination by inspection of entrails, as GOODMAN (1956) put it — if they receive heuristic stimulation from such a correspondence view of truth: discovery is simpler to countenance than creation. Platonism in mathematics will not die out simply because that view is dispensable outside of heuristics. Nor will Platonism be forced to disappear because of the appearance of contra-intuitive results in mathematics, such as RUSSELL's paradox, or even the recent independence results in set theory. For Platonism is infinitely flexible: GÖDEL foresaw the independence results and explained them by the fact that the known formalizations of set theory are simply not the "right" ones, i. e., do not correspond to what sets really are (cf. GÖDEL, 1947).

But Platonism is of no immediate use in clarifying the meaning of mathematical constructs. As a remedy to the difficulties inherent in the formalization of set theories, it is of little avail to try to keep in mind the intuitive interpretation of ε as the "membership relation" (cf. MYHILL, 1951), for it can be cogently argued that there is no exact, objective, intuitive sense of ε (cf. ULLIAN, 1969). The best one can do is to accept the fact that no one formalization can satisfy everyone's heuristic quasi-translations of set theory [34]. While a correspondence view of truth based on a flexible Platonism may provide an undeniably useful way of speaking for the communication of complex procedures or results (how many times in an introductory course in analysis is one requested to "go to infinity"?), and for the suggestive formulation of new mathematical concepts, problems, and insights (the calculus thrived on infinitesimals for over two centuries), outside of heuristics a coherence view of truth (and meaning) is certainly the clearer and more useful conception in the philosophy of mathematics.

One can even envisage research activity in mathematics without adopting a dialectical or Platonist posture. TAKEUTI (1969) illustrates this, when he suggests that we "mature our intuition" about the width of the universe of sets (i. e., about how many sets there are in the power set of any set) by investigating the mathematical implications of an exhaustive classification of widths. This is not a call to intuition based on reference to some obscure entities, but

---

34 As VAN HEIJENOORT (1967) has pointed out, the theory of sets is in a way essentially worse off in this respect than that of numbers, despite GÖDEL's incompleteness theorems, in that the notion of truth used to establish these theorems does not resist all attempts at formalization (one can formalize an ω-completeness rule). ULLIAN (1969) also suggests a similar method of intensionalizing the intuitive notion of truth in elementary number theory.

what could be properly called a pragmatic coherence approach to the heuristics of set theory. This move takes the spotlight off LAKATOS' central preoccupation, namely, how already received mathematical concepts evolve in exactness, and focuses more on the problem of how new mathematical concepts arise.

We conclude, in any case, that there is no reason to set up the coherence view of truth against the correspondence view. Much like the linguistic and referential components of meaning, these two views of truth are complementary, and this is so in mathematics as elsewhere. In factual theories, of course, the correspondence view will be appropriate outside of the heuristic domain, in dealing with the referential aspect of the theory. In fact, DAVIDSON (1967) is right in that truth shadows meaning and vice-versa: but this in no way forces commitment to a referential view either of meaning or even of truth.

2.7 Nonexistent Possibles

In explicating LEWIS' notion of L-comprehension for terms, we took, for a given theory T, the set $O_{\mathcal{M}}$ as providing the possible objects referred to by terms; and in explicating the notion of extension for terms, we took a set O_M, for some fixed model M, as providing the actual, or existing objects. LEWIS construes L-comprehensions and extensions of propositions as consisting of possible and actual worlds, and these notions we explicated by the models in $\mathcal{M}$, setting aside one particular model M as the actual world. It is easy to see that all of LEWIS' desiderata concerning L-comprehensions and possible worlds are satisfied by our explications. We compare here our formal explication of the doctrine of nonexistent possibles with those recently proposed by RESCHER (1969) and MONTAGUE (1968).

For LEWIS, that one proposition holds or not in a given world need not determine whether some other proposition holds or not of that same world; but any given proposition must either hold or not hold in any given possible world (1946, pp. 54 ff.). Under our explication, similar statements may be made concerning predicates (or formulas with free variables) and possible objects. But for RESCHER, a nonexistent possible object is identified solely by means of a *defining description,* and consequently it may be impossible to determine whether some other predicate, not involved in the definition of a certain possible object, holds or does not hold of that object (1969, p. 96). Thus RESCHER's conception of nonexistent possibles clashes with ours (and LEWIS') in that, because of the situation described above, which RESCHER calls *descriptive incompleteness,* the law of excluded middle may not apply in determining reference to possibles.

Now RESCHER insists that nonexistent possibles are determined purely linguistically, not by ostension — only existents are determined by ostension. But an immediate consequence of this position, and a consequence which RESCHER accepts, is that two possible objects will necessarily be identical if their defining descriptions are logically equivalent (ibid., p. 97). We do not feel that an explication of nonexistent possibles should satisfy this criterion of identity, for surely what is at stake is the identification of entities of some sort, not the identification of definitions. It is a trivial observation that under RESCHER's entirely linguistic explication, nonexistent possibles are limited to being one for each kind — whereas where there is one centaur we could well expect that there be two.

It seems obvious to us that what RESCHER is explicating is not possibles, but kinds of possibles. And kinds of possibles, and the questions and answers given by RESCHER concerning kinds of possibles, can be given very simple formal explication within our framework. We can take $\text{Ext}_{\mathcal{M}}(F)$ as explication for the set of objects of kind F. RESCHER's question, whether or not objects of kind F are objects of kind G, then amounts to asking whether or not one of the formulas $(F \to G)$ or $(F \to \sim G)$ is provable (if L is semantically adequate). Clearly it is possible that neither is provable, especially if the formulas F and G have no predicate letters in common (we have CRAIG's Interpolation Theorem in mind), in which case no decision concerning the applicability of G to objects of kind F is syntactically possible (except in the uninteresting case where F or G is applied or refused to all possible objects). Thus the same phenomenon of descriptive incompleteness, and the same criterion for identifying kinds of possibles, can be obtained in our explication; but unlike RESCHER's, our possibles enjoy a marked independence from syntactical determination.

One is indeed at a loss as to how to construe a Rescher-possible independently of its defining description. While what can be said syntactically of one of our possibles cannot go beyond what is syntactically already known of it and what can be deduced from such (hypothetical) knowledge, identification of our possibles is nevertheless not entirely syntactically determinable. Thus we feel our explication gives a better solution to one main problem concerning possibles, namely to devise some principle of individuation for nonexistent entities. If possible objects are to be objects at all, they must be given some freedom from syntax. Set-theoretic semantics brings to bear a second, higher level of discussion, and can therefore offer an individualization of entities which, at the syntactical level, are undiscernable.

Another major dilemma concerning possibles is how to handle at the same time existents and nonexistents. RESCHER's proposal that any predicate must either apply or not apply to an existent, while its applicability to a kind of nonexistent may be essentially syntactically undecidable, does seem more realistic than LEWIS' view. We may accomodate this double standard for reference by allowing the semantical relation ϱ_M, where M is the chosen actual world, to determine reference for existents, while for possibles in general we will allow only *relative* syntactical determination of reference, in the sense that what can be decided concerning the applicability of G to what have already been identified as F-objects will depend on the relations of entailment which can be determined between F and G. Since ϱ_M is compatible with the syntactical relations in the sense of (D2), in the case of existents (which also are possibles), ϱ_M "seconds" η much as ostension seconds definition and description of existents in natural languages. One can thus employ an interesting combination of correspondence and coherence views of truth and meaning in explicating the difference between referring to existents and referring to nonexistents.

On the other hand, RESCHER permits cohabitation of existents and nonexistents in some all-inclusive domain of individuals over which variables may indifferently run (ibid., p. 100), whereas in our explication these two kinds of possibles belong to distinct models: nonexistents inhabit possible but not actual worlds, while existents inhabit the actual one. Cohabitation of existents and nonexistents is in an obvious sense distasteful, especially as it leads RESCHER — who condemns QUINE's dictum that to be is to be the value of a variable — to institute a linguistic (syntactical) distinction between existents and nonexistents, by proposing a logic with special quantifiers to range only over existing entities and with more general quantifiers for both existents and nonexistents[35]. Yet the principle, that the distinction between what exists and what does not should not be linguistically (syntactically) determinable, seems to us to be beyond question.

Let us sum up what separates RESCHER's view of possibles from LEWIS', and our reconciliation of the two. LEWIS assumes that reference to nonexistents can be decided completely, in some absolute way: this we explicate by the usage of $\varrho_\mathcal{M}$. RESCHER wishes that reference to nonexistents be decided only relatively, and via their defining descriptions: this we explicate by the use of η, by asking whether $F\eta G$ or $F\eta(\sim G)$ holds. RESCHER finds sense only in linguistically enquiring whether reference to a kind of nonexistent holds, and allows extralinguistic determination of reference only for existents, through ostension; while LEWIS allows himself much more powerful means of deciding reference to nonexistents, through an absolute reference relation which is every bit as discerning as ostension. Our framework permits efficient explication of both views and suggests a neat compromise which incorporates the pragmatic appeal of RESCHER's view as well as LEWIS' more satisfactory ontology for possibles: reference to a nonexistent, as determinable by a speaker of L, is decided linguistically and relatively, through η, whereas identity of nonexistents is not wholly linguistically determinable but ultimately depends on the identity relation in $O_\mathcal{M}$.

MONTAGUE (1968) also condones cohabitation of existents and nonexistents, and includes in each *pragmatic language* a unary predicate symbol E to be interpreted as "exists". Any *possible interpretation* of a pragmatic language, in MONTAGUE's sense, comes equipped with a universe U of *possible objects*. U behaves in the way expected of a universe for a model, and E singles out in each possible interpretation a subset of possible objects to be taken as existing. Thus MONTAGUE's explication of possibles also violates the principle that what exists should not be linguistically determinable.

The feature which distinguishes an interpretation of one of MONTAGUE's pragmatic languages from ordinary interpretations is that the former furnishes a set I of *points of reference*. An element of I might be regarded as a moment of time (possible moment of utterance), or as an ordered pair combining a person and a moment

---

35 RESCHER only proposes this logic for "heuristic purposes" (ibid., p. 101).

of time. The extensions of formulas in the language are allowed to vary in the possible interpretations according as the point of reference varies. Thus while a certain person at one moment may have in mind a certain set of possible objects as existing, he may alter the extensions of E in the various possible interpretations at another time.

We cannot raise the same objections as above to MONTAGUE's use of an existence predicate and to his allowing existents and non-existents to cohabitate in the same universe, since he is dealing with pragmatic languages and with what some person may say exists, not with what actually exists. We do find his presentation complex, however, and puzzling. Is a given person at a given time supposed to have in mind, or to otherwise determine, what objects he is considering as existing among all the possible objects in all the possible intepretations of the language? And who is determining what interpretations are possible to begin with? MONTAGUE does not say. It is already unacceptable, *pragmatically* speaking, to assume that in a single possible interpretation, the person at a given time has a precise idea of what objects exist; recall our discussion of vagueness in Section 1.8.

Our proposal on nonexistent possibles can handily be adapted to a pragmatic context as follows. We retain the device of a set of reference points, but simply require that at a given moment a given individual should indicate which model he considers as actual. This is in keeping with our practice above of not permitting existents and nonexistents to cohabitate one and the same universe; that is, we reject a Platonist view of existence, in a sense to be described in Section 3.1. This also respects the principle of not countenancing one and the same object as inhabiting several different worlds, that is, we also reject what we shall refer to in Section 3.1 as the variant Platonist view which allows the transfer of existence through different situations; MONTAGUE's speaker has no hold on reference, hence no means of distinguishing between an object as element of one possible universe from that same object as element of another universe, and so would presumably attribute the same standing of existence to that object throughout such universes.

In our proposal, an object is inviolably linked to the interpretation in which it lies: the interpretation determines referentially all the qualities of the object, i. e., all the formulas which hold of it. But the question of what exists is an extra-syntactical one, not to be

answered linguistically. Set-theoretic semantics, in our formal explication, models ostension. The pointing out by a subject of a particular interpretation as the actual one, serves to indicate that this
way of interpreting what is said is, in his estimation, the most appropriate way of doubling what is said syntactically by adding a referential representation of what are the facts.

Our solution has the merit of dovetailing naturally with the long-
standing practice (cf. LEWIS) of distinguishing between actual and
possible worlds. In MONTAGUE's pragmatics, the speaker apparently has no say as to what worlds are possible, nor, therefore, as
to what objects are possible. By utilizing the set of reference points
outside of the interpretations, rather than inside, we feel that a
simpler and more correct pragmatics is obtained. The speaker then
may also single out some proper subset of interpretations from the
class of all logically possible interpretations, as constituting the class
of *possible* interpretations. This offers a very simple way of distinguishing between the usual notion of logical entailment (logical
meaning) and a properly pragmatic notion of entailment (pragmatic
meaning) based on the entailment relation induced from reference
to these possible models. Thus the pragmatic entailment relation
may vary according to point of reference, i. e., according to speaker,
time, etc. This explication is very easy to reconcile with Kripkean
semantics for modal logics, and offers as much power for analysis
as MONTAGUE's.

But if one wishes to be really pragmatic, say about belief sentences, then neither our apparatus nor MONTAGUE's can explain
how a given person's beliefs may go against logic. One could still
explain Jones' belief that $9 \neq 3^2$ in a "logical" way, by saying that
Jones does not have in mind all the postulates of arithmetic, or
even has in mind some false ones, so that the range of his possible
interpretations goes beyond or to one side of the logical possibi-
lities: Jones could still in this way be considered as logical with
himself. But many excellent mathematicians, having the correct
axioms of Euclidean geometry firmly in mind, nevertheless believed
just as firmly that the parallel postulate could be proved from the
others. To assume the human subject steadfastly "logical" or consistent with himself seems to us blatantly anti-pragmatic. As we
pointed out in discussing GOGUEN's explication of inexact concepts
in Section 1.8, in dealing in pragmatics we must above all guard
ourselves from complacency.

3. Existence

3.1 The Thesis that Existence is Consistency

In discussing the existence of mathematical entities, it is convenient to distinguish two points of view. To entertain a certain community of existence between abstract objects such as relations, qualities, numbers, and so on, and real objects such as chairs and apples, is to adhere to what we shall call *Platonism*. Often the common type of being which Platonists ascribe to all objects, abstract or not, is distinguished from that of actual things: they then speak of "subsistence" *(Ansichbestand)* (cf. VON FREYTAG-LÖRINGHOFF, 1951), reserving "existence" for actuals, and claim that subsistence is the ontological essence of real things, which possess in addition other characteristics such as spatio-temporality. We will call *Aristotelianism* the opposite view, that there is no type of being common to objects of different categories: such objects may perhaps exist in related or analogous ways, but there is no kind of existence which can serve as common denominator for all types of objects. These appellations will be useful in bringing out the flavour of some theories to be examined below.

The thesis that the question of the existence of mathematical entities is reducible to that of their consistency was strongly voiced at the turn of the century by POINCARÉ (cf. BETH, 1966, p. 70 and p. 642), when CANTOR's reportedly absolute Platonism came to grief on the paradoxes. The doctrine that consistency is an essential attribute of existing things has been around at least since ARISTOTLE, and surfaces continually in the literature. We agree entirely with NAGEL's argument against interpreting as ontological truths such logical principles as the principle of non-contradiction (NAGEL, 1944). Inconsistency can only be located in discourse, among statements, and not in the world, among existing things. Our fondest wishes and most consecrated practices notwithstanding, there is simply no clear grounding for the view that logical principles express the limiting and necessary structures of the world. The proper grounding of logical principles is not to be sought in their conformity with an absolute structure of facts, but in their normative role in instituting appropriate linguistic usage. The principle of non-contradiction is a regulative principle of language.

However correct this analysis may be, the fact remains that consistency has firmly embedded itself in that part of the mathematician's vocabulary with which he discusses ontological questions. This can be taken as a thinly disguised Platonist move. Before the advent of the modern explication of consistency as a property of sentences or of sets of sentences, the claim that consistency is an essential (indeed, the *only* essential) property of existing things may have been interpretable as what we have called an Aristotelian tenet, in so far as each different object could be construed as possessing its own brand of consistency, distinct from that of its fellows. In fact, one often receives the impression, when reading early writings on the subject, that the consistency of an existing object surges up from the wells of its being whenever the existence of the object is put to the test as to whether a given attribute is applicable to it or not. But with the explication of consistency obtained through the formalization of mathematics, consistency definitely becomes a property which is transferable, through relative consistency proofs, and hence, as an ontological principle, fits a Platonist view.

Thus while POINCARÉ and others may have proposed the thesis that existence in mathematics is nothing more than consistency — let us refer to this as the *EC thesis* — as a reaction against Platonism, we see that this thesis can easily be interpreted as a fresh brand of Platonism, a restricted Platonism in the sense that no such common property as subsistence is accorded to all beings, both real and non-real[36], but also a quite absolute Platonism in the sense

---

36 The manner in which VON FREYTAG-LÖRINGHOFF attributes "subsistence" to all such beings is already obscure: "we must deal with all non-Reality (or abstract Reality) in this way ... we must ascribe to it, independent Subsistence-in-itself *(Ansichbestand).* And while in the case

that the mode of existence of the various mathematical objects is uniformized. Aside from this Platonist rendering, the *EC* thesis can also be interpreted as a *boutade,* as a flat denial that the question of the existence of mathematical entities makes any objective extra-linguistic sense. We would adhere to the thesis if this were its content, but this is not the way in which it is generally interpreted.

Let us look at the *EC* thesis more closely. As enunciated by CANTOR himself, the thesis presents the existence of a mathematical object as a consequence of the consistency of its properties: "Mathematics is entirely free in its development, and is bound only in the self-evident respect that its concepts are non-contradictory in themselves, and also stand in fixed relations, ordered by definitions, with those concepts formed beforehand, already present and tried ... As soon as a number [or any other concept] satisfies all these conditions, it may and must be considered as existing and real in mathematics ... For the correct formation of a concept, the process is, in my opinion, always the same: one takes a thing without properties, which is at first nothing but a name or a sign *A,* and one gives it in an orderly fashion several, even infinitely many intelligible predicates, the sense of which is known from already given notions, and which must not contradict one another ... when one has brought this process fully to an end, all the conditions are present for the awakening of the concept *A,* which slumbered within us, and it enters ready-made into existence, provided with the intrasubjective reality which can only be required of all concepts ... to determine its transcendental signification is then the matter of metaphysics" (1883, sect. 8, pp. 182 ff.).

In the Aristotelian tradition, the relation between the consistency of a set of predicates and the existence of objects satisfying

----

of concrete Reality, this is real, in the case of ... abstract Reality, it is fictitious ... this fictitious Being only appears to be absolute in character within a ... restricted framework of other thoughts ... namely, only so long as it encounters no logical contradictions ... as soon as it becomes confronted with its first contradiction ... its unequivocal nature which is the essential prerequisite of every Being-in-itself *(Ansichsein)* breaks down, and it stands both logically and ontologically condemned" (1951, p. 24). From what we can make of this, it would seem that subsistence is attributed to all Beings but can be withdrawn as soon as a Being is recognized as being a non-Being ...

these predicates is commonly considered to warrant deducing the consistency of the predicates from the non-emptiness of the intersection of their extensions, or deducing the emptiness of this intersection from the demonstrated inconsistency of the predicates. Both deductions are in the same direction, from existence to consistency, the second type of deduction being the contrapositive of the first: consistency is thus held to be necessary to existence.

CANTOR and the *EC* thesis posit more, namely that consistency is also sufficient for existence in mathematics. But this is to take one's fancies for facts, and also to mistake one's intuitions for exact concepts. The set of all cardinal numbers seems just as unsuspicious as the set of all natural numbers, yet proves faulty: did it enjoy a kind of provisional existence before being found inconsistent? The same obscure give-and-take of existence is evidenced in VON FREYTAG-LÖRINGHOFF's explanation (cf. fn. 36).

The *EC* thesis remained unconvincing on the point of how existence is to be deduced from consistency, until GÖDEL's Completeness Theorem confirmed the believers in their faith, that "we cannot talk consistently without talking about something". We have already remarked upon such uncritical attribution of ontological import to model theory in Section 2.1, and will expand on this in Section 3.7. For the moment, let it suffice to point out that the most popular proof of the Completeness Theorem at present (HENKIN's) essentially consists in constructing a set-theoretic model for a consistent theory out of the signs of the theory itself: so much for its ontological import. The Completeness Theorem has no more than linguistic content. It establishes a relation between a syntax based on recursive arithmetic and a semantics based on set theory, a relation, that is, between what is nothing more than two (different) ways of speaking. If the *EC* thesis is truly vindicated in this way, then the thesis is indeed to be taken as a *boutade,* as affirming that mathematical "objects" only have a linguistic "existence".

The modern vindication of the *EC* thesis is thus a hoax, and the thesis remains wishful thinking. Even under the fiction of some sound relation between existence and consistency in the context of formal theories, the result, due to the impossibility of obtaining, in general, anything more than relative proofs of consistency, would be a domino theory of being, which might be amusing, but hardly enlightening.

3.2 Empiricist Notions of Existence

Having found no more than verbal justification for the *EC* thesis, we reject it, on the grounds that whatever existence may be, it should not be entirely linguistically determinable (recall Section 2.7). Rather than explicate existence at the theoretical level, several mathematicians and philosophers have sought to explain mathematical existence through appeal to the applications, or to certain aspects of the practice, of mathematics. The examination of some of these empiricist views will eventually suggest what we deem to be a reasonable solution to the question of mathematical existence.

For example, the present efforts to develop the "predicative" and "constructive" approaches to the foundations of mathematics, which are largely due to the persistent intuitionist critique of mathematical practice, might suggest considering mathematical objects as existing only in so far as they can be accomodated in a constructivist theory. But even if the debate over what is constructive in mathematics were at last resolved, there would remain the problem of explaining in what sense constructions exist (how does an assemblage of existing entities exist?) and of developing an ontology for the preexisting entities which serve as building-blocks for the constructions (the mystical faith of the early intuitionists in the natural numbers will not do).

However, no mathematician has as yet indicated a rational approach to these problems; only the interesting line of enquiry opened by the psychologist PIAGET (BETH and PIAGET, 1961, pp. 319 ff.) is sufficiently developed to permit even tentative evaluation. In point of fact, present-day mathematical constructivists are more interested in guaranteeing the consistency of their constructions and in demonstrating their fruitfulness than in debating the mode of existence of, say, infinite collections of objects, even though this ontological question was one of the principal preoccupations of the early intuitionists.

CARNAP (1950) disposes of the problem of the existence of mathematical or of any other kind of entities by distinguishing between external and internal questions of existence. Those existence problems formulated within the framework of a language are *internal* problems, while those concerning the existence as a whole of the system of entities referred to by the language are called *external*. Depending on whether one is dealing with a formal or a factual theory, internal questions of existence may be dealt with by logical or by empirical methods respectively.

Thus, presumably, the question of whether a certain mathematical construct exists within a given theory is equivalent to deciding whether a certain existentially quantified formula is provable or not in that theory. On the other hand, for CARNAP, the external question, whether the system of objects referred to by a linguistic framework exists, is simply nonsensical. The only questions of external nature which it makes sense to pose are questions of efficiency, fruitfulness, simplicity, and the like, and answers to these questions serve in determining whether or not to adopt a certain linguistic framework, or way of speaking. This external view parallels that of NAGEL concerning the role of consistency as a normative principle regulating linguistic usage in the pursuit of human goals[37].

Limiting ourselves to an aspect of CARNAP's proposal which has not been brought out in its discussion, it seems to us that, applied to factual theories, CARNAP's analysis is self-defeating, in that whether or not one takes a realist, a subjectivist, or other attitude before the real world, necessarily colours the decision as to what linguistic forms, or norms, one finds efficient or fruitful.

Upon reflection, the same can be said of mathematical theories. For instance, the present interest in constructivist mathematics would never have come about without the dogged perseverance of a handful of visionaries (the early intuitionists) in sticking to a metaphysical ideal, however justified they might have been in so doing. This is a clear case where an ontological stance (refusal of impredicative definitions, worship of the integers and of constructibility) influenced the rejection of one linguistic framework and incited the development of another. Certain ontological beliefs can be mirrored in language, can be at least partially captured in linguistic forms, and hence can be validated on the very basis which CARNAP proposes, namely the fruitfulness of the linguistic framework which they encourage. It can be convenient and useful to speak, or to refuse to speak, as if something exists. Hence external questions of existence can make some sense, and even internal problems of existence can be permeable to this sense — intuitionism delimited its own logical methods.

---

37 These pragmatic criteria are also emphasized by GOODSTEIN (1968), ROBINSON (1966b), and, most poignantly, by COHEN (1971).

It can also be doubted whether isolating internal questions of existence in mathematics from external questions is reasonable from another angle: the internal question, whether a proof of $(\exists\, x)\, F(x)$ is obtainable or not, has no ontological import in itself, without adding, say, that the existential quantifier is to be interpreted existentially, as stating that "there is (exists) an object x such that F applies to x", and not, say, substitutionally, as "some substitution instance of F is true" (cf. MARCUS, 1962). Certain translations, or rather quasi-translations in the sense of Section 2.3, must be made from internal to external contexts, if there is to be any objective ground at all for interpreting a proof as having some ontological import. Again, we stick to the principle that what exists is not to be linguistically decidable.

Hence in mathematics one can arrive quite easily at a conclusion diametrically opposed to CARNAP's, that is, that internal questions of existence make no sense whereas external questions do. But the best counsel is no doubt to consider internal and external questions as interrelated: questions of existence are generally neither raised nor answered in a single one of these spheres alone[38].

GONSETH once construed logic as a physics of "arbitrary objects", a view which he obtained "by abstracting away from all determinate qualities, from any mode of precise [spatio-temporal] localization, etc., and by retaining only the fact of existence or of non-existence, and eventually the fact of being able to possess certain attributes and of being able to belong to certain classes. *The end result of this abstraction* [process] *is what we call the idea of the arbitrary object*" (1936, p. 3). This view clearly hints at attributing an ontological quality to sets, and seems to be motivated by the naive theory that logico-mathematical entities are abstracted directly from real objects. While it is tempting to take this position on mathematical theories which are close to their applications, it is not applicable to many theories which have no such obvious links with reality.

Later GONSETH developed his conception of mathematics as the dialectic of the ideal with the empirical, presenting the infinitesimal

---

38 QUINE's views are usually discussed in parallel with CARNAP's, e. g., QUINE (1951) does envisage a transfer of linguistic ontological commitment through various interpretative levels, but we will discuss QUINE's ontological theories in detail in Chapter 4.

calculus, for example, as "a dialectic of the continuum", as a mental structure which represents a certain approximate perception of reality. More generally: "Mathematics initiates and brings about an arbitrated dialogue between a developing consciousness and a knowledge being constituted . . . in order that what is real should become, in our minds, what it is perhaps appropriate to call a horizon of reality [*un horizon de réalité*]" (1948, p. 48).

But even this more flexible explication more properly concerns applied mathematics, and factual science generally, and sheds no light on the origin and nature of an ontology for that considerable part of mathematics which, to all appearances, thrives in total disconnection from reality. While GONSETH's generous view of the dynamic interplay between the rational and the empirical phases of scientific activity provides a more satisfactory framework for the explication of ontological questions in the factual sciences than did CARNAP's dichotomy, it is not applicable without modification to non-factual theories.

BERNAYS (1950) has sought to apply GONSETH's views to pure mathematics by invoking a "mathematical factuality" to play the role of reality in GONSETH's picture of factual science. He considers a mathematical entity, the existence of which is asserted (no doubt through the existential interpretation of the quantifiers) by some theorem in a formal theory, to enjoy a relative mode of existence, or *bezogene Existenz,* with respect to the formal theory. This relative existence refers not so much to the particular formal theory, but rather to the "total mathematical structure", or *Gesamtstruktur,* having been formalized: we would say that the relative existence of a construct in a formal theory is relative to the natural mathematical theory formalized.

Ultimately BERNAYS must elucidate the mode of existence of such mathematical structures. This he does by referring further to a *Gedankensystem,* or conceptual framework: "We finally end up with the reference to an ideal framework. This is a conceptual system which involves a kind of methodical attitude, to which the mathematical existence-propositions lastly refer . . . the mathematician moves with assurance in this ideal setting, and has at his disposal here a kind of *acquired evidence*" (1950, p. 19, our underlining).

Such a high degree of certainty fades, however, as one leaves the well-tested, classical core of mathematics, and heads for the foundations. BERNAYS meets this objection with GONSETH's theory of knowledge, following which in mathematics, as in all other sciences, knowledge is not absolute, but progresses through a continuing dialectic between the rational and the empirical.

BERNAYS sees the undisputed, consistent-by-experience, classical core of mathematics, or *mathematische Tatsächlichkeit,* as playing a role analogous to that of the actual world in, say, theoretical physics. As example of this mathematical factuality he gives the laws of Euclidean geometry, which remain independent of its different axiomatizations, or the properties of the continuum, which are independent of the manner in which the reals are introduced: "These relations, to which we are, so to speak, forcedly led as soon as we admit certain types of calculations and of mathematical operations, play the role of a datum for foundations research, and which [datum] is to be theoretically determined more precisely" (ibid., p. 21).

To distinguish his position from Platonism, BERNAYS emphasizes that mathematical factuality is not made up of objects of the type of real things, but of relations and structural connections: "There is no question here of a being, but of relational, structural bonds and of the emergence (by induction) of ideal objects out of other such objects" (ibid., p. 23).

We are thus somewhat left in the dark, finally, as to what properly ontological status applies to the *Tatsächlichkeit* or to the mathematical objects which enjoy *bezogene Existenz.* BERNAYS wishes to avoid Platonism, yet seems willing to entertain his structural connections and relations as existing in some sense, since he speaks of them as "objects". Though we find the explicit and essential relation between internal and external problems of existence in BERNAYS' conception to be a valuable correction of CARNAP's absolute dichotomy, there seems to be little else behind BERNAYS' ontological vocabulary than a positive attitude towards the objectivity of mathematical activity.

Returning recently to this problem, BERNAYS finds that mathematical objectivity rests on the objective *sui generis* reality of schemas, and characterizes mathematics as "the theory of schemas, according to their intrinsic nature" (1970, p. 57), as "the science of idealized structures" (ibid., p. 65), a *schema* being an idealized, approximate representation of a part of reality. Mathematical objectivity arises from the phenomenal objectivity of the structural aspect of reality, through the processes of idealization and abstraction: "In mathematics we have not mainly to do with structures given in a directly phenomenal way, but with idealized structures, where the idealization consists in an adaptation to conceptuality, in a compromise, so to speak, between perception and conceptuality" (ibid., p. 58). Having thus emphasized how mathematics inherits, or induces, a kind of objectivity from the real, BERNAYS then turns to describing how mathematics consolidates this objectivity through, in particular, axiomatization.

Once again we are left to feel that the apple has passed by just out of reach. In consolation we might decide that BERNAYS has led us about as far as we can go in the direction of an empirically based ontology for mathematics. We can decide that rather than persist in seeking to ascribe some mode of existence to mathematical structures, a temptation which to date has led to no tangible result whatever, it is preferable to turn resolutely away from the barren — if not false — problem of mathematical "existence", and to direct our attention to the much more promising problem of explaining the phenomenon of objectivity in mathematics. The difference in perspective so obtained will allow us a greater mobility in shifting the spotlight from referentially inspired structures to the structuring of mathematics through entailment relations, a mobility which seems to us essential if we are to explain the continued development of mathematical theories long after the referentially perceived structures which first prompted them have entirely faded from sight.

3.3 Objectivity and Evidence

The problem is to explain how there can be objectivity without appeal to objects, how there can be a mathematical factuality which does not appeal in any direct way to reality. What founds the certainty and clarity which permeates mathematical activity? But first of all, in what does this objectivity consist?

After having frankly sized up the case for the notion of mathematical object, KREISEL, for one, seems to favour the move to objectivity (1958, p. 138). One of the main themes in KREISEL's philosophical writings is that of *informal rigour:* "Formal rigour does *not* apply to the discovery or choice of formal rules nor of notions ... The 'old fashioned' idea is that one obtains rules and defini-

tions by analyzing intuitive notions and putting down their properties ... the 'old fashioned' idea assumes ... quite simply that the
intuitive notions are *significant*, be it in the external world or in
thought ... Informal rigour wants ... to make this analysis as
precise as possible ... in particular to eliminate doubtful properties of the intuitive notions when drawing conclusions about them,
and ... to extend their analysis, in particular, not to leave undecided questions which can be decided by full use of evident properties of these intuitive notions" (1967a, p. 138).

In this conception of informal rigour, Kreisel sets considerable
store by intuition, to such a point that Moss (1971) persistently —
and, we feel, mistakenly — associates his position with realism,
i. e., that there are mathematical objects and abstract structures.
This sort of assertion even to-day does not seem to interest Kreisel
(cf. 1967b, p. 221, and 1970, pp. 20 ff.): he explicitly insists that
his use of "object" in the locution "mathematical object" is only
meant to indicate "objectivity" (1970, p. 35). Kreisel's attitude here
is very close to that of Bernays. Both intend, through the phrase
"mathematical object", to point towards the patent phenomenon
of objectivity in mathematics, rather than to indicate any commitment to Platonism, which both consider to be a superfluous hypothesis.

Kreisel's notion of informal rigour, however, touches upon two
sorts of objectivity, that founded on the phenomenon of evidence,
and that which derives from the intuitive analysis of mathematical
concepts such as set or number. We consider that it is preferable
to discuss the objectivity of mathematics in relation to the phenomenon of evidence alone. Kreisel's description of the mathematician's experience of evidence: "He reflects on his argument and sees
if he understands it" (ibid., p. 27), is as impregnable as it is un-
dramatic. But it is misleading to lump together this sort of objectivity with that arising from Kreisel's "conceptual analysis", which,
through its apparent appeal to some objective source of truth, can
indeed be mistaken for a welcoming gesture to realism.

Putnam also links up objectivity with truth in a similar way:
"The issue of the 'existence' of 'mathematical entities' must be sepa-
rated from the question of the objectivity of mathematical statements (e. g., do all well-formed formulas of set-theory have a truth-
value?). If one wishes to regard set-theory as 'objective' (all well-

formed formulas have a truth-value, independent of provability), *how* much set-theory? All of *ZF*? Just the theory of types? ... Just analysis? Just predicative analysis? etc." (1968, p. 284).

But surely such "objectivity" is *relative* to one's conception of "the truth". The constructivist's "objective" core of, say, set-theory, will be different from the category-theorist's. The mathematical factuality of a given theory remains relative to the particular intuitive perspective taken on it. KREISEL himself follows up his assertion that "every significant piece of mathematics has a solid mathematical core ... and if we look honestly, we shall see it" (1958, p. 158), with an honest avowal that what is seen as mathematical core can vary with the viewer. This being said, we hasten to agree that the phenomenon of evidence in mathematics derives in part from Kreiselian intuitive analysis, in as much as the constant sharpening of mathematical concepts is part of the latter practice.

KREISEL's and PUTNAM's willingness to make room for truth in mathematics is by way of reacting to the absolute conventionalist "if — then" of formalism. But the recent proposal which KREISEL makes concerning how to objectively decide the truth of, say, the continuum hypothesis in set theory (1970, pp. 38 ff.), is also an "if — then" situation, albeit an informal one. Any broadening of the accepted characteristics of a given mathematical concept is always a matter of choice: it can be argued that some choices are "objectively" more desirable than others, but such argument always rests on some premise. What KREISEL's proposal shows is that there *can* be some such argument for the acceptance of certain undecided propositions of set theory, which is an advancement at the present stage of the game. But there is no reason to assume that hard conceptual analysis, or informal rigour, will determine the characteristics of a mathematical concept in an absolute (objective) way: the proliferation of distinct set theories remains a possibility. The objectivity of the "intuitive" notion of set is, at the least, not evident.

Refusal to entertain a deterministic interpretation of mathematical objectivity is in harmony with the relativization of the phenomenon of evidence. As BERNAYS (1946) has pointed out, mathematical evidence is, in fact, *acquired*. The mathematician understands his arguments in a particular conceptual context (we would say heuristic *milieu*), and another mathematician, once immersed in this *milieu,* can share his understanding. This objectivity of mathematical arguments, relative to a certain context, is the basis for the community of understanding (MYHILL, 1951), for the *Tatsächlichkeit* peculiar to mathematics. As KREISEL (1967a) amply underlines, in the various crises in mathematics it is either the axioms (e. g., the comprehension axiom) or the rules of inference (e. g., the law of excluded middle) which are at stake, not the deductions themselves. Even LAKATOS' very sceptical case study of exactness in mathematics (1963) bears this out.

We readily echo KREISEL, then, in rejecting the formalist theme, that the formal level is the only meeting-ground for objectivity in mathematics. Formalization is undisputably an excellent *maieutic* for the explanation of natural mathematical theories, for the explication of mathematical constructs, and for the elucidation of the notion of meaning itself, as we have seen; but it holds no absolute monopoly on mathematical objectivity, and indeed would be impossible without informal analysis. In particular, the sweeping claim that formal proofs provide a higher degree of evidence than informal deductions collapses under the most elementary analysis (KREISEL, 1970, pp. 22 ff.).

We do not, then, expect the informal, intuitive analysis of mathematical notions to lead in a single, necessary direction, and rather consider the objectivity of mathematics to reside in its deductive organization, in the evidence derived from proof. It is proof that cements communication in mathematics: the reflex to rely on deductive evidence is the mathematician's effective mode of defence against subjectivity. But, as against the formalist view, such evidence is acquired, and there is no proof, no evidence, without a sufficient understanding of the concepts in play. Or, as THOM puts it, rigour is *local:* "There is no rigorous definition of rigour ... a proof is rigorous if, in any sufficiently instructed and prepared reader, it creates a state of evidence which imposes conviction" (1970, p. 230).

Having delimited in what we consider mathematical objectivity to reside, the superficially paradoxical problem remains, then, to explain this mathematical phenomenon of objectivity through local, relative evidence: how is acquired objectivity possible? An answer which nourishes itself on the very root of the paradox is given by PIAGET's genetic epistemology.

3.4 A Seasoned Constructivism: Piaget's Genetic Epistemology

BETH, through his "Fundamental Criterion of Demonstrative Force": "An argument has deductive force if it admits no counterexample" (BETH and PIAGET, 1961, p. 79), and in his subsequent detailed and very instructive discussion of mathematical evidence, emphasizes, like THOM, the vital importance of experience, or "preparation", in mathematical reasoning. This preparation of evidence in a theory can be stimulated by requirements arising from several different domains at once: through intuitive (e. g., geometrical) data, from the needs of certain fields of application of the theory, by logical criteria, through problems arising from already established mathematical theories, and so on. Unlike THOM, however, who too readily accepts the easy epistemological refuge provided by Platonism, BETH draws the general conclusion that with increasing complexity and abstraction of mathematical theories, the role of intuition in establishing evidence gradually diminishes.

PIAGET (ibid., pp. 176 ff.) takes a decisive step towards a satisfactory explanation of the phenomenon of mathematical evidence, by observing (not *positing*) that evidence develops in parallel with the emergence of mathematical "structures", that is, with the recognition of abstract relations independently of the particular "objects" between which the relations hold. Evidence matures with the progressive acquisition of structures, with the increasing objectification of the components of these structures, with the growing awareness of the autonomy of the operations performed on these components relative to the particular "objects" which at first are considered to constitute them, these "objects" themselves being structures already previously elaborated at a lower level of conceptual organization. Definite acquisition of evidence, or "noetic" integration of the result of an experience, is associated with the completion, or "closure", of the corresponding structure.

We will not give a thorough exposition of PIAGET's theory of knowledge here. The child's progressive differentiation of his first logico-mathematical experiences in coordinating his actions on objects, his gradual converting of these experiences into interiorized operations through the process of "reflective abstraction" [39], the con-

---

39 The process of *abstraction réfléchissante* consists of the reconstruction, at a higher level of abstraction, of a previously acquired structure,

tinually evolving deductive organization of mathematical concepts, the restricted freedom of the mathematician in the elaboration of his subject, all these themes are carefully developed and their validity argued in detail by PIAGET (ibid.; for a very condensed but more up-to-date presentation, cf. 1970a or 1970b). There is no more passionate yet painstaking exposition of how, in mathematics, existence yields to deduction, of how objectivity is progressively freed from the object, and of how, to use KREISEL's words, proof is derived from meaning; and there is no more illuminating criticism of Platonism, realism and intuitionism in mathematics. We limit ourselves here to a few comments on PIAGET's constructivism.

One of PIAGET's most striking observations is his verification of the total absence of differentiation between subject and object in the very young child (1969, p. 127; he accordingly considers FREUD's theme of narcissism, at this early stage, to be a *"narcissisme sans Narcisse"*). Here indeed is a leading argument in the debate over correspondence and coherence views of truth and meaning, and in that over the possibility of a radical distinction between analytic and synthetic judgements. It obviously establishes the necessity of incorporating a diachronic dimension to any epistemology which aspires to reasonable completeness. Objectivity and evidence are, in a surprisingly strong sense, progressively acquired.

PIAGET's genetic approach, which treads a nice middle route through the Scylla and Charybdis of whether new mathematical "facts" are invented or discovered, renders Platonism in a very definite sense unnecessary, by providing an adequate account of the foundation of mathematical knowledge, and, at the same time, explaining satisfactorily how new concepts and theories arise. PIAGET shows how there can be objectivity without existence, and new objectivity without pre-existence.

The mathematician need not be reduced, then, to asserting that "mathematical structures exist independently of the human mind

---

the previous construction being integrated into the higher-level structure. Examples in mathematics are the axiomatization, say, of Euclidean geometry, where a "naive" theory is treated to a more formal, deductive formulation, or the construction of the rational numbers from the natural numbers, where the latter are perceived as a part of a more perfect whole, which enjoys "closure" with respect to certain operations. A category-theoretic formulation of a fragment of mathematics is certainly a fine example of reflecting abstraction.

which thinks them ... their type of existence is no doubt different from the concrete, material existence of the external world, but is nevertheless subtly and profoundly tied to objective existence ... how else can we explain their decisive success in describing the universe?" (THOM, 1970, p. 229). For, from PIAGET's perspective, mathematical structures can very well get along without existence, though they are related, as constructions perhaps several times removed, from the real universe which is their genetic base. Unless we wish to take up again the needless metaphor of *bezogene Existenz,* or to float a similarly bogus concept of, say, "derived existence", there is no necessity to ascribe to mathematical "objects" any kind of existence at all.

PIAGET's account allows us to both make this ontological economy and still whole-heartedly agree with THOM that "the important mathematical structures (algebraic, topological) appear as data fundamentally imposed by the external world, and ... their irrational diversity finds its only justification in reality" (1970, p. 136). PIAGET and INHELDER even go so far as to view the linking of biology to mathematics as "no theorist's luxury but the psychologist's duty" (1969, p. 131); "when one considers the organism's behaviour on all the levels (including the highest cognitive levels), one finds the connection between the subject (product of the organism) and the objects (indissoluble sectors of its "environment"): no subject without action on the objects and no objects without a structuration contributed by the subject ... in the strictest sense of the word ... knowledge is a special case of biological adaptation" (ibid., p. 157).

The genetic viewpoint not only provides a refreshing explanation of the perplexing combination of newness and necessity which accompanies the results of mathematical research, and of the astonishingly successful participation of mathematics in describing reality, but it can also serve in explaining the limitative results obtained in mathematical logic, in so far as the latter may be construed as a faithful formulation of the conceptual processes of higher mathematics. For in considering formalization as a highly refined form of "reflecting abstraction" (1961, p. 273), it is in direct keeping with the fundamental openness of the genetic, constructive approach (compare GONSETH's primal theme, *l'ouverture à l'expérience*) that each conceptual structure calls for a more highly organized one to answer to its eventually manifest deficiencies.

The contrast with the positivist position on mathematics, that with increasing rigour, content decreases, is obvious. In PIAGET's constructivism, meaning, or content, is cumulative and the evolution of mathematical constructions is towards ever-increasing comprehensiveness and rigour. That two such contrary horses can be teamed follows from the fact that logico-mathematical constructions ever build on those which came before, integrating the latter while surmounting some of their inadequacies. There is no irreconcilable conflict here as there can be with factual theories based on reality; in mathematics, what corresponds to the Popperian derring-do of advancing perilous — but, precisely for this reason, highly significant — hypotheses, is the imaginative implementation of sophisticated reflective abstractions. Very correctly, mathematics is recognized as an activity differing fundamentally from factual science[40].

In this light, too, Platonism, at least in the form of the EC thesis of Section 3.1, is relegated to its proper rank of impotent daydreaming. For the establishment of the existence of mathematical entities through proof of the consistency of the structures containing them meets with the insurmountable difficulties expressed in GÖDEL's results. On the contrary, it is instructive to re-examine, from PIAGET's constructivist perspective, CANTOR's descriptions of the "discovery" of mathematical entities (cf. in particular the quotation in Section 3.1): they then appear as overly colourful narrations of the peculiar "closure" phenomenon accompanying the labours of reflective abstraction qua an objective taking-into-account, a conscious *realization,* at a higher level of conceptual awareness, of the coherence and completeness of a particular constellation of previously acquired constructs[41].

It is not less enthralling to thus picture mathematics as marching towards continually perfected objectivity, instead of towards un-

---

40 It is disappointing that SCHEFFLER (1967) still sees fit to deal with "Science and Subjectivity", "Meaning and Objectivity" and "Epistemology of Objectivity" without mentioning Piaget's work, critically or otherwise — or again, asserts that "for the purposes of mathematics and science, it is sameness of reference that is of interest rather than synonymy [sameness of meaning]" (ibid., p. 57).

41 Though conceptually more demanding, PIAGET's constructivism certainly throws more light on what is going on in LAKATOS (1963) than does the lazy man's solution, Platonism.

attainable objects. It is more correct to envisage objectivity in this way as a *process,* and not as a state. The continual meeting of the meanings of mathematical concepts in the minds and manners of different mathematicians is not to be explained by seeking *out there* the "that" in the expression "means that".

3.5 Heuristics and Mathematical Existence

There is one phase of mathematical activity in which to speak of existence makes sense. That is the moment when, engaged in research, in seeking to "discover" new interrelations between the constructs holding his attention, the mathematician attributes a referential quality to those constructs.

The physicist, or any natural scientist for that matter, may at some moment commit himself temporarily to a similarly vague ontology. Whether or not a professed realist or subjectivist on other occasions, when investigating new constructs it is simplest and most efficient for him to put ontological finesse aside and adopt a wholesale Platonist attitude. In the throes of creating, the physicist finds relief in adopting, instead of the grandiose mantle of the creator, the more practical garb of the discoverer, and considers constructs such as "electron" or "electrical field" as referring, however vaguely, to actual entities.

In the same way, the mathematician may find it convenient, when squeezing properties and relations out of the construct "natural number", to visualize himself in the process of counting a countable infinity of actual objects, or, when prying out the properties of "set", to contemplate some paradise of multihued entities. Doing original research is hard enough: if the researcher feels more comfortable when deep in a plushy Platonist sofa, there is no reason to disturb him by telling him the sofa's not there.

Taking on certain ontological hypotheses is a valuable heuristic trick in the process of extracting meaning from recondite constructs. In our opinion this is the one context where the mathematician may be considered to commit himself ontologically as deeply as anyone else, and the only context where speaking of existence in mathematics literally *makes sense:* the sense which is thus made, or created, is then taken hold of and developed in a natural or formal theory. The ontological umbilical cord then becomes useless, and should be severed. That most mathematicians do not bother to do this does not invalidate this point, as most mathematicians will also agree that to understand a theory in mathematics means essentially to grasp its logical development, that is, the relations of entailment among its constructs, rather than to perceive a denotation for the constructs. Intensional relations at the theoretic level crystallize insights gained through referential quasi-translations at the heuristic level.

Heuristically, then, Platonism can be a potent enough kind of day-dreaming. The most outstanding example of this, perhaps, is the role played by the infinitesimal in the communication and development of the calculus. This referential heuristic agent evaded proper objectivization until long after its decline in the favour of analysts. Perhaps ROBINSON's capture of the infinitesimal in his non-standard analysis, through a clever employment of the logical notion of a non-standard model, is only retributive justice, a demonstration that the notion of the infinitesimal resided only in a manner of speaking, although its formalization requires an especially close examination of mathematical "speech". It does seem significant for PIAGET's theory, at any rate, that what was accomplished by WEIERSTRASS' school was an objectivization, a *prise de conscience* of the *operation* of passage to a limit which advantageously could replace the infinitesimal.

From PIAGET's viewpoint, a Platonist heuristic component to a mathematical theory would not be unacceptable, inasmuch as it would be a continuation of the original situation presiding at the inception of logico-mathematical experience, where the child first becomes conscious of the objectivity of various operations carried out on real objects. Indeed, the very language in which we carry out our research is pregnant with ontological commitment. Both QUINE (1958, p. 19) and his critic ALSTON (1958, p. 256) agree that the grammatical repertoire of European languages makes it simplest for users of those languages to speak of abstract entities as if they existed.

In a direction other than linguistic, BETH even finds that for many mathematicians Platonism expresses a "psychic reality" and that, even though heuristic appeals are almost always entirely eliminated or concealed when the results of mathematical research are presented in definitive form, "there exists a kind of "pre-established harmony" between pure mathematical thought, the deductive method and Platonism; the alliance between these three is so stable, that it is difficult to consider it as resulting only from a gratuitous historical juxtaposition" (1961, pp. 110 ff.). BETH concludes that a psychological study of the typology of mathematicians is essential before anything more objective can be said about the preferred heuristic attitudes of mathematicians.

It is no doubt partly in the name of the widespread and often useful heuristic practice of Platonism, or realism, in mathematics that KREISEL over-reacts to GÖDEL's watering down of his earlier, apparently vividly realist position by underlining the parallel possibility of adopting a frankly pragmatic approach: "there might exist axioms so abundant in their verifiable consequences, shedding so much light upon a whole field ... that, no matter whether or not they are intrinsically necessary, they would have to be accepted at least in the same sense as any well-established physical theory" (GÖDEL, 1964, p. 265).

To be sure, KREISEL's examples of the practice of informal rigour, or of "staring at concepts" (cf. LAKATOS, 1967, p. 102), all have happy endings, and his own illustrations of "philosophical" analysis are outstanding (we shall take one up in Section 4.1). But epistemologically, as a general picture of how mathematical concepts come into being and evolve, his position does not seem more enlightening than is the strictly post-mortem positivist criterion of fruitfulness.

Again, assertions such as POZGAY's (1971), that the operation which is at stake in the axiom of choice is clear to the mind, and that its completion involves only "practical" difficulties, may well fail to establish a state of evidence inspiring conviction on the part of the reader, for lack of sufficient instruction and preparation. Our intuition of the truth of the choice axiom is better mellowed through deductive experimentation with its consequences and interrelations with other suggestive principles than through Platonistic speculation. In theory, in the matter of objectivity, we prefer COHEN's self-doubting, conservative attitude (1971) to POSZGAY's "liberal intuitionism" (which could more simply be called Cantorism).

In practice, we may relax our guard and enjoy the best of both worlds, as COHEN puts it. But even in the process of mathematical creation, there is no advantage to championing the realist style over the pragmatist style, or a referential, correspondence view of truth over the deductive coherence view: recall Section 2.6, especially TAKEUTI's strategy of "maturing our intuition". Certainly the most generous attitude would be to temper d'ALEMBERT's *"Allez de l'avant et la foi vous viendra"* with a more or less strong dose of Kreiselian informal rigour, according to taste[42].

The present situation in set theory can be taken to illustrate PIAGET's genetic conception of mathematical development. The realization of certain deficiencies, e. g., the undecidability of the continuum hypothesis, in the present structure (the ZERMELO-FRAENKEL axioms) calls for a new construction to surpass these deficiencies. What orientation(s) this higher construction will take, and what reorganization of our concept(s) of "set" will be effected, is now in the brewing; the utilization of conservative extension results as in KREISEL (1970) is one possibility, category-theoretic considerations, as in FEFERMAN (1969), is another. Quite possibly no single style will predominate, and ZERMELO-FRAENKEL set theory may ultimately be recast in a new role similar to that played by the group axioms in algebra, as MOSTOWSKI (1967) surmises.

What does seem certain is that the Platonist style in the philosophy of mathematics, having been dealt serious blows by the paradoxes (cf. BERNAYS, 1935) and by the limitative results on formalisms (deflation of the *EC* thesis by GÖDEL's Second Incompleteness Theorem), can no longer count on the unconditional support of set theory. Hopefully this will permit a long-neglected popular reevaluation of realism in mathematics, which, if conducted with a minimum of critical acumen, will come to the conclusion we suggest: that the proper place for mathematical realism is in the kitchen.

3.6 Style

The sensitive point in a constructivist epistemology is to explain how the new structures emerge from the old. In his treatment of this articulation between old and new in mathematics, PIAGET, not wishing to espouse a deterministic line which raises more questions than it answers, emphasizes the vast degree of freedom which the mathematician enjoys before the moment becomes ripe for a further "reflective abstraction". On the other hand, desirous of ex-

---

42 MOSS is far off the mark when he comes to the conclusion that "it can be helpful to think *as if* there are abstract structures, only as a guide to the choice of fruitful axiom systems" (1971, p. 435). There is obviously no *a priori* guarantee of fruitfulness in a realist heuristic strategy, any more than there are axiom systems which are *a priori* fruitful.

plaining the feeling of definiteness and necessity which accompanies such a move, PIAGET depicts mathematical evolution as responding to definite "vections", or directions, which are not present or recognizable *a priori* in the old structures, but which can be identified retrospectively once the reconstruction has been carried out.

The problem then remains to describe how these vections first arise. PIAGET is silent on this point, perhaps because a glance at the history of mathematics reveals cases where the progression towards new constructions is accomplished through astonishing, if not contradictory, reversals. The case of the infinitesimal, which rendered highly stimulating intuitive service in the early development of analysis, only to be repudiated summarily at a later date, is exemplary. In contemporary mathematics, the set-theoretic approach which so effectively served to unify in a single descriptive context the apparently diverging fields of analysis, algebra, probability theory, geometry, etc., is by some mathematicians considered nowadays obsolete: their favour for such a descriptive tool goes to category theory, which owes in great part its ascendance to the very unified view of mathematics obtained through structural resemblances made evident by set theory. A "foundations" which raised the object to the rank of building-block of the mathematical universe is railed by an irreverent offspring which boasts that objects are quite dispensable (cf. MACLANE, 1968).

Every bit as abruptly, then, as in other domains (e. g., artistic, vestimentary) where man's inventiveness is primordial, style in mathematics is seemingly subject to unpredictable change. Nevertheless, style it is that prepares the ground for new mathematical constructions by prompting the exploration and sustaining the development of already existing theories. We might view style, therefore, as the source of PIAGET's "vections" in mathematics. A remarkable attempt to circumscribe this phenomenon has been made by GRANGER (1968), who illustrates at length the nature and import of successive geometrical styles which he identifies as Euclidean, Cartesian, Arguesian and vectorial.

As an objective manifestation of the rapport between theory and practice which we often emphasized in Chapter 2, as intrasubjective transmission of a particular perception of a natural mathematical theory, as heuristic communication of certain aspects of that theory which yet remain to be captured in a deductive mold, stylistic effects can hint at and help to gradually reveal much of the semantical (in the broad sense) iceberg left uncovered by the stricter deductive formulations — what we called syntactical interpretations — of a natural theory. In KREISEL's terminology, style would be the hand-maiden of informal rigour. As such, an elucidation of the concept is preemptory for the philosophy of mathematics. To this end we comment upon a few key passages from GRANGER's pioneering work.

GRANGER defines a *signification* as *"that which* results from the placing into perspective of a fact within a whole, illusory or authentic, provisional or definitive, but in any case experienced as such by a consciousness . . . the *fait de style* . . . is by its very nature inseparable from a signification; it is the significant fact *par excellence"* (op. cit., p. 11). In opposition to the "manifest, thematic structuration" carried out by science on its contents, GRANGER construes style as the "latent, experienced structuration" of scientific activity itself. Style is at once a certain manner of introducing the concepts of a theory, of linking them together, of unifying them, as well as a certain way of delimiting the intuitive contribution in the determination of these concepts.

For example, the static vectorial representation is one means of grasping the concept of complex number; a different *fait de style* is its (dynamic) representation as an operator defined on vectors. In the second case, the intuitive contribution of the geometric situation is to make the rules of complex multiplication, as a composition of transformations, appear natural; the former *fait de style* renders, in turn, the addition laws natural. The representation of complex numbers as two-by-two matrices removes all strangeness from the relation $i^2 = -1$. The quotient ring of the ring of polynomials in x with real coefficients relative to the principal ideal generated by $x^2 + 1$ offers yet a fourth, more abstract mode of representation.

"[In] these different ways of grasping a concept [that of complex number], of integrating it into a system of operations and associating intuitive implications to it . . . it is evident that the structural content of the notion is not affected in this case, that the concept, as mathematical object, subsists identically under these stylistic effects" (p. 21). GRANGER adds that this is not always so, however, and gives examples, through his description of successive geometrical styles, of stylistic posits which command veritable conceptual mutations: "in any case, the orientation of the concept changes towards this or that usage . . . style thus plays a possibly essential part both in a dialectic of the internal development of mathematics and in that of their relations with worlds containing more concrete objects" (p. 21).

The harmony between these views and those of PIAGET becomes even more striking as GRANGER progressively sharpens his tentative definition of style to: "a latent and naive structuring of the set of residues left by a certain reflected and thematic structuration of an experience" (p. 102). The structure under development does not exist as an object, but as a *perspective of an object,* the factual expression of which is a particular unity in the orientation of research moves, which are otherwise unjustified. "The structure as object *in fieri* is one, but its *signification* varies as a function of the *faits de style* ... "signification" is not to be confounded with an intrastructural *meaning* ... [it] cannot be described as a structural object ... without its losing its essential character" (pp. 103 ff.) [43]. As with PIAGET, a persistent openness, subject to continual readjustment of orientation, is fundamental to GRANGER's epistemological position.

From our own viewpoint, we may construe style in mathematics as the manifestation of significations derived from what we called the heuristic component of mathematical theories. It is through *faits de style* that heuristic clues, though inexplicit, may be transmitted within the mathematical community. The *work* of the mathematician is then to develop or even transform a mathematical theory in the direction indicated by a given style: significations (which may be partially of referential nature) are gradually transmuted into meanings, or entailment relations, between constructs. Style is thus the essential complement to reflective abstraction, as signalling the semantical residue left without articulate expression in a given *prise de conscience,* or conceptual realization.

In another sphere, PIAGET's own achievements in epistemology offer an outstanding example of a careful and faithful exploitation of the heuristic clues provided by his earlier experience in biology. PIAGET's theories are not a simple transposition of a fragment of biology into the epistemological realm; the objectivization of the insights gained from his biological perspective gives rise to properly epistemological hypotheses, followed up, wherever possible, by ap-

---

43 GRANGER's caution here contrasts sharply with LEWIS' overbold "pragmatic" treatment of the notion of signification (cf. Section 2.5).

propriate argument and verification procedures. This vein of inspiration being far from exhausted, it is to be expected that the genetic style in epistemology will continue to assert itself for some time to come.

In the philosophy of mathematics, the constructivist style appears to hold out considerable interest, as we have tried to illustrate above. As a possible outcome of its rise in strength and of the decline of Platonism, we might even witness a mutation of the way of regarding formal existence proofs, perhaps along the lines described by KREISEL (1967b, pp. 244 ff.): the preference for "constructive" existence proofs, which not only establish "existence", e. g., of a solution to a system of partial differential equations, but also identify more or less completely some of the "existing objects", is unanimous. In a rabid constructivist mood, one might be tempted to go so far as to hold that a non-constructive existence proof is of no more than heuristic value.

In mathematical practice itself, the birth, life and death of different styles, and the simultaneous co-existence of conflicting styles (e. g., fluxions and infinitesimals) within the same theory, seem to depend entirely on the ingenuity and heuristic powers of the illustrators themselves. The petering out of theoretical returns from a given style can prompt its abandonment, and even the abandonment of the theory itself, for lack of further heuristic incentives; yet neither are to be counted as definitely dead, but only as dormant, awaiting rebirth in some new mathematical context.

A determined study of the development of mathematical theories and styles may nevertheless throw some light on the phenomenon of style in mathematics, by isolating certain general tendencies in the manner of interaction — or of interference — between different styles. One of the interesting topics which could be investigated under this heading would be to explain the relatively slow progress of constructivism in foundational research until the last decade. A more ambitious project would be to undertake a broad comparative study of the logicist, formalist, etc., tendencies in foundations research at the turn of the century. Under the stylistic perspective, it is to be expected that a certain coherence and solidarity among these movements could be progressively revealed, resulting in a more cohesive picture of research efforts in this field than is usually given.

As the beacon of objectivity, as the wet-nurse of evidence, style strangely remains a theme largely ignored in mathematical philosophy. Yet we feel that recognition of the vital importance of mathematical style can help to explain and reconcile the paradoxical combination of necessity and liberty in the evolution of mathematics. Stylistic studies in mathematics would encourage the development of a healthy mathematical heuristics, and a healthy philosophy of mathematics, by illustrating BERNAYS' conclusion: "The requirement of mathematical objectivity does not preclude a certain freedom in constructing our theories. We should use this freedom to build our mathematical theories in such a way as to be, as far as possible, comfortable homes for our intellect" (1967, p. 112).

3.7 Sets and the Semantics of Mathematics

Our reasons for rejecting, in Section 2.2, the received referential view of meaning, as given by more or less formal set-theoretic interpretations of mathematical constructs and theories, can now be made more clear. For one thing, though our first experience of the set concept and of one-to-one correspondences undeniably issues from contact with real objects and operations carried out on them, as much can be said, from PIAGET's viewpoint, of any other mathematical construct. As such, sets are not privileged by some superior form of mathematical existence, i. e., objectivity. The set properties, like those of any other mathematical construct, result from successive and possibly diversely motivated objectivizations, or reflective abstractions. Sets enjoy, therefore, no special primitive status as objects of reference in mathematics.

On the other hand, sets have proven to offer an easy hold for our intuition, and accordingly the set-theoretic style has found widespread application, including set-theoretic or Tarskian "semantics". But this universality collapses when intuition seeks stimulation in divergent contexts. For instance, BOOLOS (1971) can argue that the iterative conception of set, which is itself the expression of a mild Cantorian constructivism, does not decide even the choice axiom.

From PIAGET's point of view, such independence results in set theory are not surprising. They simply establish that a certain intuition of set is not as strong as it was thought to be, and that sustained motivation for "deciding" certain set-theoretic hypotheses must be sought elsewhere, perhaps in more than one heuristic context and style; an example of how category-theoretic considerations

7*

may influence set-theoretic postulates is given in the next section. To "decide" such postulates may well require a decision of higher order, involving fundamental *revision,* or reconstruction, of the set-theoretic edifice based on a fresh reflective abstraction, on a re-orientation of objectivity.

Whether or not some unexpected, strong inspiration may appear and re-unify set theory is, from this perspective, beside the point. Set-theoretic "semantics" constitute neither a properly referential pole for mathematical meaning, nor a universal semantics in the broader non-referential sense. A set-theoretic interpretation, either semantical or syntactical, of a formal or natural mathematical theory only provides certain *faits de style* through which certain significations may be evoked, these significations being proper to set theory and perhaps not at all germane to the style (or styles) which habitually accompany the interpreted theory — this clash of styles being sometimes surprisingly fruitful.

The advent of a category-theoretic foundation for mathematics (cf. LAWVERE, 1966, or MacLANE, 1968 or 1971) illustrates this relativist position concerning the semantics of mathematics. It is possible to interpret the entire mathematical corpus in category-theoretic terms. In particular, LAWVERE (1966) has turned the tables on set theory, and has presented a formalization of category theory in which it is possible to "do" intuitive set theory; set theory can itself be carried out within category theory. Put yet another way, LAWVERE has "reduced" sets to categories. This is sufficient to ensure that all of mathematics can be "founded" on category-theoretic notions.

But we would not wish to take this talk of reduction and of foundation literally. Just as set-theoretic formulations of various parts of mathematics do not fully capture the variegated heuristic components of the different theories, category-theoretic formulations emphasize certain "structural" features of the same (cf. the following section) to the detriment of other, perhaps more "native", semantical aspects. We have criticized elsewhere (1972) the over-enthusiastic usage of the epithet "natural" in describing a successful category-theoretic formulation of a fragment of mathematics. Not only is such "naturalness" relative, it may also be a misnomer for what category-theorists wish it to express, which seems more akin to "suggestiveness", a quality we shall be concerned with in Sections 4.2 and 4.3.

Maintaining that there is just one "semantics" for mathematics is incorrect; maintaining that it would be best that there be only one is bad heuristic strategy. BERNAYS (1935) already intimated as much. In terms of the distinction made at the outset in Section 3.1, we prefer an Aristotelian, as opposed to Platonist, conception of mathematical "existence", i. e. heuristics. To this we would add that mathematical semantics are not only diverse, but are also in constant flux, through the rise, interaction and decline of variously successful heuristic styles. Mathematicians build on shifting foundations.

As an illustration of the possible cross-fertilization between different semantics for mathematics — a "foundation" for mathematics is such a semantics, in that its ultimate goal is to give a particular definite meaning to mathematics — we can take the set-theoretic problems connected with certain constructions of categories from categories. We will describe one such category-theoretic construction in the next section; the relevant point for the moment is that various such constructions are permitted or not according to the set-theoretic foundational system used in setting up the theory of categories. Arguments based on the naturality and desirability of certain category-theoretic constructions may influence a decision as to which axioms to adopt in order to extend a given set theory, in that the possibility of effecting these constructions imposes certain demands on that set theory if it is to be used in modelling categories. FEFERMAN (1969) contains an up-to-date discussion of set-theoretic foundations for category theory.

Paralleling this consideration is the relativity of the notion of "the" category of sets to the set-theoretic system which furnishes the models (in the model-theoretic sense) for the theory in question. The divergence of different possible set theories on the question of the feasability of certain infinitistic constructions has real significance in this context, and obliges one to relativize the notion of "the" category of sets to the system of ZERMELO-FRAENKEL, or to that of GÖDEL-BERNAYS, etc. One accordingly obtains several distinct categories of sets (for details cf. HATCHER, 1968).

More generally, it is to be expected that category-theoretic considerations will somewhat influence the ascendancy of various set-theoretic styles in so far as a given set theory proves more germane or useful in category-theoretic contexts; and *vice-versa*. Both category theory and set theory may be used to describe the same mathematical construct from different viewpoints, that of set theory being, intuitively speaking, biased towards substance, that of category theory being biased towards structure. But the very fact that both approaches provide foundational systems for mathematics obliges each to take the other into account. Set-theoretic formulation of a new mathematical construct will require that an effort be made to give a categorial formulation of the construct, if category theory is to remain competitive, and conversely. The category- and set-theoretic styles, or ways of speaking, have their distinct advantages, and can be expected to amply exemplify the phenomenon of cross-fertilization of languages and theories which will be taken up in the next chapter.

Category theory is not only a new foundational competitor for set theory, joining in this respect such old hands as intuitionism; it also presents "natural" situations where the apparent monolithism of set theory breaks down. It vigorously illustrates the actual relativity of mathematical semantics and the cross-fertilization of these different semantics, as well as providing a highly suggestive constructivist view of mathematics in which objects appear only in an initial and eventually dispensable capacity. As such it deserves a closer look on our part.

3.8 Categories and the De-ontologization of Mathematics

The importance of the advent of category theory for the problems of meaning and existence in mathematics lies primarily in its availability as an alternate foundation for mathematics with a primitive conceptual basis quite alien to that of set theory. The theory of categories is tailored to handle structural similarities and dissimilarities in mathematical constructs and theories. In contrast to the notions of natural number, real number or set, which carry a certain ontological aura inasmuch as they are habitually conceived as arising by abstraction from objects or from collections of objects, the notion of a category can be described as a technical concept developed to study the various structure-preserving correspondences which abound in mathematics. For instance, categorial[44] methods provide a very satisfactory description of the phenomenon of the

---

44 We coin the term "categorial" to replace the cumbersome "category-theoretic" and to allay confusion with "categorical", in the sense of unique-up-to-isomorphism.

"natural" isomorphism between a finite-dimensional vectorspace and its double dual.

In so far as mathematics is centrally concerned with such structure-preserving correspondences, category theory is a viable replacement for set theory as a unifying and explanatory foundation. This abstract bias of category theory, its disdain of talk of objects — "Even in foundations, not substance but invariant form is the carrier of the relevant mathematical information" (MacLane, 1968, p. 293) — consummates the de-ontologization of mathematics which was practically completed in the last century by the proliferation of various "reductions" to one another of apparently distinct mathematical theories. Set theory, the last half-serious prop for speaking of mathematical existence which remained after these reductions, is seen to be dispensable. Ironically, set theory itself, through the unified view it gave of heuristically disparate theories, was an essential tool in the development of the contemporary structural approach to mathematics which finds category theory so congenial, and which indeed prepared its conception.

Let us informally introduce, in a slightly unorthodox way, a few categorial notions before further discussing some of the stylistic particularities of category theory.

A category is made up of a collection of *objects* and a collection [45] of *morphisms*. A morphism may be considered to relate one object to another in the sense of indicating some (possibly very weak) similarity between the two objects. For example, the perfect similarity between an object and itself is consecrated by the categorial postulate that for each object in a category, there is a special morphism, called the *identity morphism* of that object, which relates that object to itself. Those morphisms which indicate perfect similarity without necessarily entailing identity are called *isomorphisms*.

Just as it is natural to consider that similarity between objects, in some fixed sense of "similarity", is a transitive relation, so is there a categorial postulate which permits the unique composition of morphisms under suitable conditions. It is then further postulated that composition of an identity morphism with another morphism always yields nothing else but the latter morphism. This postulate serves to characterize identity morphisms. Isomorphisms can also be

---

45 This apparent appeal to set theory is obviated in a formalization of category theory (cf. Lawvere, 1966).

characterized through sole use of the notion of composition of morphisms.

For example, in the category the objects of which are sets and the morphisms of which are ordinary set-theoretic functions, the isomorphisms are bijections or one-one-onto functions; for a category with groups as objects and group homomorphisms as morphisms, the isomorphisms will be ordinary group isomorphisms. Objects in a category are distinguished by the morphisms which do or do not hold between them. Whatever one can "say" categorially — that is, literally, say with morphisms — of a given object, one can likewise say of any other object isomorphic to it; for example again take sets of identical cardinality, or isomorphic groups.

The all-important step in setting up a category with a certain type of object is the decision as to what kind of relation or correspondence should be taken as a typical morphism. The aim, intuitively, is to judiciously choose as morphisms those correspondences between objects which will permit the satisfactory description of all of the structural properties judged to be intrinsic or fundamental to the kind of object under study, i. e., the aim is to be able to "say it with morphisms". Consequently, it is not surprising that category theorists find their constructions and results "natural". Emphasis on the relational or structural aspect of mathematics is so absolute in category theory, that the first thing a category theorist tends to suggest after the intuitive introduction of the basic notions is to "throw away the objects" (e. g., FREYD, 1965, p. 108). Since everything that one can say with morphisms about an object can be said about its identity morphism, the very notion of object may appear redundant; but in practice, the categorist is not so inflexible.

Categories themselves can easily be construed as objects in a larger category, the appropriate type of morphism between categories being that of a *functor,* which is essentially a function from the collection of morphisms of one category to the morphisms of another category which "represents", or preserves, what can be said with morphisms in the first category; a functor is, therefore, a kind of meaning-preserving correspondence between categories. To thus consider categories themselves as objects in some larger category is an example of what we called in the preceding section a construction of categories from categories. There are more complex constructions of categories from categories which arise quite naturally, but we cannot go into greater detail here (cf. HATCHER, 1968).

That objects in a category be distinguished — and ultimately identified — by the morphisms holding between them, is most fitting if one takes the position that mathematical thought construes its objects as susceptible of differentiation, and grasps them, so to speak, by means of their differences. GRANGER finds this view explicitly stated in GRASSMANN (1844) and develops it in an interesting analysis of the "vectorial" style in geometry, whereby the notion of magnitude is dissociated from that of geometric entities (1968, pp. 92 ff.); this suggests that a parallel may be drawn here with the categorial style, which tends to dissociate, more definitely than does set theory, the notion of property or relation from the notion of mathematical object.

Together with MÖBIUS' barycentric calculus and the well-known *Erlanger Programm* of KLEIN (according to which geometry is the study of those properties of geometric entities which are invariant with respect to a given class of geometric mappings) and its obvious generalizations, these vectorial and categorial styles are positive manifestations of a more general and more diffuse tendency, characteristic of modern mathematics, towards a de-ontologization of the discipline. Category theory is itself thus seen as constituting in its own right a *fait de style,* as expression of a "philosophical" style deriving from the steady ascension and realization of what we termed in Section 3.4 a constructivist view of mathematics.

It is to be expected, then, that in this context categorial formulations of various parts of mathematics will be found excitingly "natural": category theory is simply a very adroit illustration of this general epistemological style. The *suggestiveness* of a categorial formulation of, say, the concept of number, flows in large part from the newness of this de-ontologizing view in contrast to the traditional, more customary view of the numbers as counting sets[46]. As we pointed out elsewhere (1972), the suggestiveness of a particular approach to a given mathematical construct or theory is indeed a superficially paradoxical compound of naturalness and unnatural-

---

46 In particular, it makes quite unnecessary BENACERRAF's lengthy discussion (1965) to the effect that numbers need not be identified with sets. LAWVERE has given an admirable categorial formulation of the PEANO system, consisting of a single transparent axiom (cf. MACLANE and BIRKHOFF, 1967, p. 67).

ness: the former in the sense of the faithfulness and efficiency of the approach as expression of the particular heuristic view taken, the latter in the sense of the newness of this strongly expressed heuristics in contrast with the traditional formulation and heuristics of the mathematical fragment.

One must accordingly exercise discernment in interpreting the tenet, often voiced in mathematical folklore, that category theory had its humble beginnings in efforts to describe the phenomenon of "naturality" in mathematics. This naturality actually consists in an evidence of similarity in mathematical structures, including mathematical proofs, in so far as they too are constructions. The categorial approach, with its emphasis on "naturality", must not be interpreted, therefore, as expressing a new kind of realism in mathematics.

Nor, to be sure, should category theory be thought to provide a "blanket view" of mathematics, any more than does set theory. For one thing, only the diversity of heuristic *milieux* can explain the presence of distinct mathematical theories (upon which presence category theory was dependent for its very conception!). For another, sustained appeal must often be made to these diverse *milieux* in seeking to verify that a given categorial formulation of a certain theory is correct (the categorial formulation must respect the theory!) or in striving to prove a new fact concerning the theory and inspired by categorial considerations; categorial formulations can suggest directions in which our knowledge of a given theory may develop, but such knowledge most usually is acquired only after dirtying one's hands and doing some "hard" mathematics.

The categorial talk of naturality expresses, then, a *parti-pris* for structural considerations within and between mathematical theories. The apparent abuse of the epithet "natural" in actual informal categorial discussion also attests to the very high degree of adequacy attained by category theory in the expression and application of this structuralist option. The heuristic power of the categorial approach is such that admonitions to "write down the evident diagram, apply the obvious argument, obtain the usual result, and, above all, do what comes naturally" already figure in the mathematical humourist's repertoire. "Natural" is thus used as synonym for "obvious-from-the-categorial-perspective", and its over-use is a patent evocation of a community of understanding essential to this particular type of mathematical evidence: a community of understanding at least as sound as that underlying set theory, and which, through its successful use in describing the numerous structural interconnections discovered in modern mathematics, inspires at least as much confidence in its consistency.

A mild ontological twist to the categorial style is worthy of comment here, as it exemplifies how convenient and common it is to slip into an informal ontological way of speaking in the practice of mathematics, however vibrantly one champions structure over substance at the philosophical level of the evaluation of the finished product. The reader will recall BETH's acknowledgement (Section 3.5) of a "pre-established harmony" between pure mathematical thought, the deductive method, and Platonism.

In category theory a peculiar *diagram notation* is used, whereby composition of morphisms is exhibited through displays of interconnected arrows resembling illustrations of oriented graphs in graph theory. This notation has become indispensable for the practical communication and development of category theory, relieving the researcher of the rigours of thinking in terms of intricate compositions of morphisms. This is accomplished so completely and efficiently, that the diagrams themselves become objects of study: the categorist easily swings into talk of "chasing around diagrams", "embedding one diagram in another", etc., so much so that the practice of category theory reduces in large part to constructing or "drawing" the right kind of diagram (called *commutative* diagrams). It is a novel twist, which brings home the extreme unpredictability of mathematical ingenuity, that in the heuristics of category theory the notation itself in this way becomes reified[47].

---

47 We permit ourselves here a light remark on the so-called picture theory of language, inspired by this categorial diagram notation. Without attempting to link our remarks with any particular picture theory, we simply indicate a sense in which reference can be considered to literally "picture" the relations of entailment in a formal language or theory.

In Proposition 5 we obtained a dual homomorphism from $\mathscr{A}^k$ to $\mathscr{E}^k$, that is, a structure-inverting correspondence from intensions of formulas of rank k onto their extensions. If the Tarskian definition of extension is used, restriction to rank k can be dropped, and a dual homomorphism from the full calculus of intensions $\mathscr{A}$ to the full calculus of extensions $\mathscr{E}$ is obtained. In either case, one has essentially two Boolean (even cylindrical) algebras, and a dual Boolean homomorphism between them.

Now, every Boolean algebra is a preordered set, and every preordered set can be viewed in a natural way as a category, with the elements of the

Indeed category theory does seem to present, within the sphere of mathematical practice, a powerful combination of abstract reasoning and spatial intuition which accords well with KANT's epistemological doctrines (cf. BETH, 1961, pp. 11 ff.). Category theory enables the mathematician to formulate theorems in such a way that their proofs "leap to the eye", as KREISEL puts it (1967b, p. 210). A less perfect instrument with which to "show" and to "see" mathematical theorems (etymologically, as THOM has emphasized, a theorem is the object of a *vision*) would make much more problematic, if not impossible, the grasping of those highly sophisticated connections which category theory handles with relative ease.

In another sphere, there are embryonic applications of category theory to epistemology which are of interest, and which we mention in closing this section.

By giving a categorial formulation of the dual situation which holds between finitely axiomatizable theories and the classes of all their set-theoretic models (a situation described in Section 1.6 as a referential one, where (D2) and its converse hold, that is, where intensions are comprehensions), LAWVERE (1969) proposes that a start is thus obtained at describing the relation between what he

---

underlying set as objects and with the preorder relation supplying the morphisms (there will be at most one morphism between any two objects). A Boolean homomorphism between two preordered sets, considered in this way as categories, is nothing else than a functor between these categories. Thus the reference relation can be viewed as giving rise to a dual or *contravariant* functor from the category of intensions, where the morphisms are relations of entailment between intensions, to the category of extensions, where the morphisms are inclusions. Diagrams in the first category may be considered as presenting relations between meanings, and diagrams in the second category as representing relations between reference classes: the dual functor of reference must literally respect, or preserve, the meaning diagrams, save for inverting them.

Put another way, a reference relation must reflect meaning configurations: in the category of reference classes, these configurations will be reflected with morphisms inverted, as in a mirror. In the case of a categorical theory, reference does nothing more than mirror syntactical meaning relations, *and conversely:* the reference functor in this case has what is called an *adjoint* functor (in dealing with preordered sets, it is more common to describe such a situation as a *Galois correspondence;* cf. BIRKHOFF, 1966). In the case of noncategorical theories, the distinct reference functors ϱ_M add further relations, particular to the different models, to those mirrored from the category of intensions.

calls the Formal and the Conceptual aspects of mathematical activity. Without defining what he means by "Conceptual", he does give examples of what he considers to constitute Conceptual activities in mathematics, to wit, to "visualize geometrically" the curve corresponding to a polynomial equation, or to consider classes of "actual" groups to which theorems of group theory "refer" (ibid., p. 281). As "Formal" activities he considers the algebraic manipulation of the aforesaid polynomial equation, or the deduction of theorems from the axioms of group theory.

LAWVERE's proposal of identifying, on the one hand, the Conceptual aspect of *Foundations,* which he defines as "the study of what is universal in mathematics" (p. 281), with categories of a certain general sort, and, on the other hand, the Formal aspect of Foundations with Logic and eventually with formalized languages and categories (again "of a certain sort"), would seem to stand or fall according to whether all mental activities relative to mathematics admit description in categorial terms. But LAWVERE's proposal is presented so vaguely (from the examples given, Formal operations seem to be no more than particularly limited Conceptual ones) that all that one can say about it, at this stage, is that it seems to express no more than a wildly nebulous brand of neo-Pythagoreanism stemming from a category theorist's enthusiasm for his subject [48].

Not every mathematician active in foundations research is as cautious as BERNAYS, who opines that, especially in the light of the GÖDEL and TARSKI incompleteness results concerning a given theory's essential inadequacy for the formulation of its own semantics, mathematics as a whole is not itself a structure, or mathematical object, nor is it isomorphic to one (1970, p. 64; compare also THOM, 1970).

LAWVERE's dreams of empire appear difficult to reconcile with PIAGET's constructivism, with the open-ended philosophy of mathe-

---

48 In continually admitting to finding difficulty in distinguishing standard models from the non-standard, the father of non-standard analysis has likewise made himself suspect in KREISEL's eyes: "For those of us who love the sinner even when we abhor the sin, it is interesting to ask what is behind ROBINSON's tactics. Could it be that some universal principle of indefiniteness of basic, standard concepts strikes him as a plug for the philosophical importance of non-standard models in place of standard concepts?" (KREISEL, 1969, p. 110.)

matics which we have found so stimulating above. It seems to us
that the mere observation of the aforementioned duality between
axiomatizable theories and their classes of models hardly shines
forth as "an essential feature of any attempt to formalize Foundations" (ibid., p. 282). A description of the dynamic interaction, or
cross-fertilization, of the Formal and the Conceptual in a particular
context (to begin with, in category theory itself, as we have sketch-
ed above) gives a more faithful and interesting account of what
transpires in foundations work, than does their simple separation
through a statically conceived state of duality. That such a fluctuating phenomenon is circumscribable — let alone formalizable — is
far from evident[49].

More sound and more promising is the possibility, advanced by
LAWVERE, that categorial generalization of such dual situations in
mathematics may furnish insight into which directions future mathematical research may fruitfully take. In this heuristic vein, the ease
with which dualistic and, more generally, hierarchical situations
lend themselves to categorial formulation strongly hints at the possibility of utilizing category theory in formulating, say, PIAGET's constructivist epistemology. Of course, we are not suggesting that the
one example which we have given above of a construction of a new
category from a class of given categories is appropriate to this end,
for its deterministic, automatic nature is alien to the openness essential to the Piagetian approach: but there is a wealth of such possible
constructions, and a given reflective abstraction may well find appropriate formulation in one of them (in a somewhat different context, categorial formulations of the basic concepts of systems theory
have been well received and offer decided advantages over set-
theoretic formulations of the same, principally on account of the
facility with which hierarchies of systems can be handled categorially; cf. GOGUEN, 1970).

In a very profound sense, then, it can be expected that category
theory will continue to erode the superficial ontological views of
mathematics encouraged willy-nilly by set theory, and to prove itself
a strong competitor for set theory in mathematical semantics and
epistemology.

---

49 GONSETH decades ago developed to a fine art the dialectics of such
dual situations (cf. GONSETH, 1970).

4. Reduction

4.1 Reduction in Mathematics

There is heated debate at the moment among philosophers of science over the concepts of reduction and inter-theoretic explanation in the physical sciences. The emphasis put by some philosophers, in particular by NAGEL (1961), on the preservation of the entailment relations in a reduction situation has been contested by FEYERABEND (1962), and by many others since then. The storm would somewhat abate if in the physical sciences a distinction were made and kept between a logical and more or less formal notion of reduction, construed as a content- or meaning-preserving relation between theories, and a less easily circumscribed notion of inter-theoretic explanation, construed as a relation between theories following which the content of the "reducing" theory serves to elucidate the reasons why the content of the "reduced" theory is not in fact realized.

Indeed, as GLYMOUR (1970) has pointed out, most cases of "reduction" in the physical sciences relate together inconsistent theories, and fall in this latter category of inter-theoretic explanations: reduction in NAGEL's sense is only very rarely exemplified in the factual sciences. GLYMOUR suggests that since explanation of a theory is not usually effected by showing why the theory is true, but rather by arguing that the theory is false because such-and-such is the case, or by showing why the theory would be true if such-and-such were the case, it would be more fitting to view the problem of inter-theoretic explanation as an exercise in the presentation of counterfactuals.

However this may be in the physical sciences, it might be expected that in mathematics the concept of reduction would be eminently transparent; but this is in fact not so. We will attempt in this chapter to clarify the notion of reduction in mathematics in the light of our preceding discussions.

Accompanying many rebuttals to NAGEL's scheme is the qualification that, in mathematics, reduction may be satisfactorily analyzed by NAGEL's main condition on reductions, namely, that the principles of the reduced theory be logical consequences of the reducing theory. This widely held view can be shown to be superficial in mathematics too, for reasons which are best seen in a more formal setting, where NAGEL's condition is appropriately explicated by the notion of syntactical interpretation.

We recall from Section 2.3 that a formal theory T' is said to interpret *syntactically* another formal theory T'', both theories being expressed in the same language L, if the non-logical constants of T'' can be replaced by expressions of T' in such a way that upon such replacement the theorems of T'' are transformed into theorems of T'. This implies, among other things, that the image of a conjunction of two formulas under such a transformation is the conjunction of their images, that the image of the negation of a formula is the negation of the image of that formula, that the image of an existentially quantified formula is the similarly existentially quantified image of that formula, and so on; and that the image of an axiom is a theorem.

But this formal notion of syntactical interpretation is already not general enough, in that it cannot be applied to mathematical systems of different logical character. For instance, it can neither account for the simple reduction of the predicate calculus to the propositional calculus obtained by dropping quantifiers from formulas to obtain their "associated statement forms" (cf. HATCHER, 1968, p. 37), nor can it account for GÖDEL's two-way reduction between classical arithmetic and intuitionistic arithmetic (cf. KLEENE, 1952, pp. 492 ff.); in the first case because of a linguistic discrepancy, in the second because of deeper logical differences.

A generalization of the notion of syntactical interpretation is that of a *translation,* due to WANG. Roughly, a translation of T'' into T' (these theories possibly expressed now in languages of different logical character) is a correspondence between T'' and T' which is formally proven to be essentially a syntactical interpretation through a normalizing appeal to some background theory T. Usually T is assumed to be a theory dealing with the natural numbers, and the proof that the interpretation is syntactical is carried out through the usual arithmetization in T of the syntactical features of T' and T''.

Somewhat more precisely, suppose B_1 and B_2 are the proof predicates of the theories T' and T'' respectively, that is, B_1 is a binary predicate in T such that $B_1(m, n)$ holds if and only if m is the number of a proof in T' of a formula in T' with number n, and similarly for B_2 with respect to T''. A recursive function f in T is called a T-*translation* of T'' into T' if f maps numbers of formulas of T'' onto numbers of formulas of T' and if

$$\vdash_T ((\exists x)\, B_2\,(x, n) \to (\exists x)\, B_1\,(x, f\,(n)))$$

for all n which are numbers of formulas of T'' [50].

Since translation in this sense requires that formal proof be given of the preservation of entailment relations, that T'' is translatable into T' appears to be a stronger assertion than that T'' is syntactically interpretable into T'. However, in so far as a syntactical interpretation of T'' in T' is usually established in some background theory T which lends itself easily to the appropriate formalization, we may view the notion of translation as a generalization of the notion of syntactical interpretation.

In fact, the usual syntactical interpretations of the non-Euclidean geometries in the Euclidean, of arithmetic in set theory, and of the complex numbers in the reals, can all be raised to the rank of translations (cf. WANG, 1963). The fact that in translating we have at our disposal a background theory containing arithmetic with which to compare, through arithmetization of their syntax, the entailment relations of theories formalized in logical systems of divergent character, permits the trivial reduction of the predicate calculus to the propositional, and GÖDEL's reduction of classical arithmetic to the intuitionistic, to be also recognized as translations (cf. KREISEL, 1955).

There is a famous completeness theorem which admits simple formulation in terms of translation, and which we take the opportunity to introduce now in anticipation of its relevance to the discussion in the later sections of this chapter. It is well known that

---

50 Compatibility of f with the logical operations rounds out the usual definition of T-translation, but we dispense with the details here (cf. WANG, 1963, pp. 353 ff.).

any consistent theory T'' formalized in the predicate calculus has a denumerable *semantical* interpretation (recall Section 2.3), or as is often said, has a "true" interpretation, or model, "in the domain of natural numbers". The question naturally arises whether this semantical interpretation of T'' in a countable universe can be "lifted" to a syntactical interpretation of T'' in classical arithmetic. Suggestively: can the truth of the axioms of T'', established in a countable universe by a correspondence approach, be expressed and established through a coherence approach, by provability in a formal number theory?

The answer is a qualified affirmative. Let T' be a finitely axiomatized theory containing the system Z of classical arithmetic (the axioms of Z are the PEANO axioms plus the recursive equations defining addition and multiplication), and let Con (T'') be an arithmetical formulation in T' of the consistency of T'' satisfying the conditions of GÖDEL's Second Incompleteness Theorem. The result is: if Con (T'') is provable in T', then T'' can be syntactically interpreted in T'. KREISEL suggestively calls this the LÖWENHEIM-SKOLEM-GÖDEL-BERNAYS Theorem (1955, p. 29), which we abbreviate to the *LSGB Theorem*. Slightly reformulated, with T' now taken to be simply $Z_{T''}$, that is, Z plus Con (T'') as an additional axiom, we get:

$$T'' \text{ is } Z\text{-translatable in } Z_{T''}.$$

This formulation WANG (op. cit.) calls BERNAYS' *Lemma*. These notions and results will be useful in discussing the "ontological" repercussions of the LÖWENHEIM-SKOLEM Theorem, but let us return now to the problem at hand.

The gain in generality over syntactical interpretation notwithstanding, there are objective grounds for finding the notion of translation unsatisfactory as explicatum for the notion of reduction in mathematics. For example, suppose T is a common *extension* of T'' and T' in the sense of TARSKI-MOSTOWSKI-ROBINSON (1953, p. 11), e. g., T may be $T'' + T'$ in the sense of Section 1.6. Suppose further that T'' is T-translatable into T' and, speaking informally, that the formula F_2 of T'' is translated into the formula F_1 of T' (more formally, a translation maps numbers of formulas onto numbers of formulas). Continuing to speak informally, the definition of a T-translation allows us to assert that $\vdash_T (F_2 \to F_1)$, but not necessarily that $\vdash_T (F_1 \to F_2)$, that is, the translation of a formula is possibly weaker, in a proof-theoretic sense, than the original formula. As KREISEL puts it, F_1 may not express the full proof-theoretic content of F_2 (1955, p. 32). The trivial translation of the predicate calculus into the propositional calculus through the "associated statement form" reduction is an obvious example of this.

A similar phenomenon is already observable in the notion of syntactic interpretation. Suppose that T'' is syntactically interpreted in T', and that formula F_2 of T'' is mapped onto formula F_1 of T': we have that if $\vdash_{T''} F_2$, then $\vdash_{T'} F_1$, but the converse may not hold. The very simple reason for this is that in a syntactic interpretation situation, proof in T' may well be easier than proof in T'', in that not all theorems of T' need be images of theorems in T'', i. e., T' may have "additional" axioms. Thus KREISEL's objection to translation as explicatum for reduction can be applied equally well to syntactical interpretation: proof in T' need not tell us anything about proof in T''. Or, as we put it in Section 2.3, intensions of formulas may be bloated under translation or under syntactical interpretation, so that meanings in T'' are not properly reflected by meanings in T'. We will expand on this formulation in the next section.

Taking as paradigm the reduction of the predicate calculus to the propositional calculus expressed in HERBRAND's Theorem, where each of the reducing formulas entails the formula reduced, KREISEL (1955) formulated the notion of what we will call an *adequate translation*. Roughly, an adequate translation of T'' into T' determines effectively for each formula F_2 of T'' a sequence (F_1^n) of formulas of T' in such a way that

$$\vdash_{T''} F_2 \text{ if and only if } \vdash_{T'} F_1^n \text{ for at least one } n.$$

Aside from HERBRAND's Theorem, the desirability of considering the truth of a formula $(\exists x) F(x)$ in arithmetic to be equivalent to the truth of at least one of the formulas $F(0), F(1), F(2), \ldots$ can further motivate consideration of such translations of formulas into sequences of formulas. KREISEL has used this strong notion of interpretation to give new and exciting formulations of HILBERT's programme of reducing "infinitistic" proof methods to "finitist" methods (cf. KREISEL, 1953 or 1964).

For an explication of formal reduction suitable for our purpose here, we need only observe that the particular case of adequate translation of formulas into constant sequences of formulas gives a fair formulation of the notion of reduction of proof-theoretic meaning. For essentially we then have that if $\mathbf{T}''$ is adequately translated into $\mathbf{T}'$, with the formula F_2 being mapped onto the sequence $(F_1{}^n)$, $F_1{}^n$ being identical to the same formula F_1 for all n, then

$$\vdash_{T''} F_2 \text{ if and only if } \vdash_{T'} F_1.$$

This is the notion of *faithful translation* which was introduced in Section 2.3.

Thus, if by "reduction" we understand a content- or meaning-preserving inter-theoretic correspondence, then the notions of syntactical interpretation or of translation are too weak to give a satisfactory elucidation of such reduction even in the extremely rarefied meaning atmosphere of formal mathematical theories. As remarked in 2.3, the essential conservative touch of non-creativity, without which a reduction is not worthy of its name, is assured by the notion of faithful translation.

The general acceptance among philosophers of science of NAGEL's condition as an adequate analysis of the "logical" reduction of one theory to another is, as a result, puzzling. The fact pointed out by KREISEL is that syntactical interpretations or translations only preserve formal entailment relations in a unidirectional sense, and that the proof-theoretic meanings of the constructs of the original theory may be distorted under such interpretations. Since it seems to be an obvious desideratum of the general notion of logical reduction that at least the intensional component of meanings be preserved intact, to the extent that our explication in Section 2.3 of the intensional component of meaning in natural (informal) theories is acceptable, the notion of faithful translation, diluted from KREISEL's too strongly mathematical notion of adequate translation, stands forth as a superior explicatum for logical reduction in scientific theories generally.

Perhaps the fact that informal syntactical interpretation has for so long been considered as the backbone of reduction and of meaning-preserving correspondence merely illustrates the lag between mathematics and the physical sciences. But upon reflexion, the reason for the inadequacy of syntactical interpretation as explicatum for meaning-preserving correspondence is seen to be more deeply seated. Syntactical interpretations, or models, were developed in the context of a problem different from that of meaning, namely, the consistency problem for mathematical theories. The establishment of a syntactical interpretation is sufficient demonstration, for example, of the consistency of non-Euclidean geometry relative to Euclidean geometry. In contrast to this, one theory will be faithfully translatable into another only if both theories are consistent (or inconsistent). The failure of syntactical interpretation to handle the meaning problem is therefore not surprising.

One can go farther yet in narrowing down the possible explications of the notion of reduction. Aside from preserving intact the meanings of the reduced theory, by the very etymology of the word one could expect that the reducing theory T' be "smaller" or "simpler" than the theory being reduced, T''. This desideratum can be naturally rendered as requiring that the reducing theory T' be a *subtheory* of the reduced theory T''. As explication of the notion of reduction we then obtain the notion of a *faithful translation into a subtheory*. This explication of reduction is doubly sound in that meanings in formal theories — proof-theoretic intensions — are most meticulously preserved through faithful translation, while the subtheory condition adds a further conservative touch.

In a more strongly mathematical vein, KREISEL proposes adequate translation into a subtheory as explicatum for formal reduction, and also adds the nice stipulation that the demonstration of the adequacy of the translation be carried out in a metatheory no stronger than the reducing theory itself (1964, p. 171). This should satisfy even the most scrupulous.

If one wishes to entertain reduction in mathematics otherwise than as proof-theoretic reduction, one must first be clear on what it is that is being reduced. Syntactical interpretation is appropriate for the consistency question, faithful (or adequate) translation seems satisfactory for meaning. Set-theoretic interpretations may also qualify as interpreting the meaning (not the ontology!) of the constructs of a formal theory in a broadened context (recall Section 2.4), so that certain kinds of model-preserving inter-theoretic correspondences may be recognized as meaning reductions in some natural way; we will mention one such type of reduction in the next section.

But if one wishes to entertain *ontological* reduction in mathematics, then one should be careful, beforehand, to independently identify "ontologies" for mathematical constructs and theories, and to justify one's use of the word. From our findings in Chapter 3, it would be surprising were we to wax enthusiastic over talk of this sort of reduction in mathematics.

4.2 Meaning-preserving Correspondences

It has become a commonplace that modern mathematics deals not with objects, but with structures. But if this saying banishes the old problem of mathematical existence, it also calls for an explanation of what is meant by "structure". Obviously one cannot mean something objectively existing, or some configuration of some kind of "objects", as this would contradict the rebuff to ontology. However sorely tempted would be the set theorist to think of structure in terms of hierarchies of "sets" and of relations between them, or the group theorist to construe the same in terms of classes of "groups" and of homomorphisms between them, or the category theorist to see structure as realized in his peculiar "diagrams", a general answer must rise above the particularities of the heuristic components of the different mathematical theories, while being applicable to each of them; and all that appears to remain as the objective common part of all mathematics is the deductive organization of its subjects at various conceptual and linguistic levels.

One commonplace is thus answered with another. But we will try to exploit positively this second commonplace, by further clarifying the notions of meaning and of meaning-preserving correspondence in mathematics, whereas the original rejection of objects in favour of structure is usually employed negatively, in consecrating the de-ontologization of mathematics. Indeed, a general conclusion to be drawn from Chapters 2 and 3 is that though mathematics is barren of ontology, it is pregnant with meaning. Meaning is discernable within formal theories by means of syntactical or semantical entailment relations, within natural theories through entailment seconded by informal rigour, and among formal and natural theories and their heuristic components through semantical and syntactical interpretations, translations, reductions, and all manner of more vague quasi-translations. It is necessary to look even more closely at the latter inter-theoretic meaning relations.

Let us first make quite clear our statements made in Sections 2.3 and 4.1, to the effect that neither syntactic interpretation nor translation (in WANG's sense) of a formal theory T'' in another formal theory $\mathbf{T}'$ guarantees that no relations of entailment can hold in $\mathbf{T}'$ between translations (i. e., images) of formulas $\mathbf{T}'''$ which do not already hold in $\mathbf{T}''$ between the translated formulas (i. e., the pre-images), whereas faithful translation does carry such guarantee[51].

Let F_2 and G_2 be formulas of $\mathbf{T}''$, and suppose they are mapped respectively by an interpretation or a translation onto formulas F_1 and G_1 of $\mathbf{T}'$. Then the formula $(F_2 \to G_2)$ in $\mathbf{T}''$ must be mapped onto $(F_1 \to G_1)$ in $\mathbf{T}'$, but under syntactical interpretation or (WANG) translation we have only that

$$\text{if } \vdash_{T''} (F_2 \to G_2), \text{ then } \vdash_{T'} (F_1 \to G_1),$$

whereas under faithful translation we have

$$\vdash_{T''} (F_2 \to G_2) \text{ if and only if } \vdash_{T'} (F_1 \to G_1).$$

So under faithful translation, Int (F_1), taken relative to the entailment relation in $\mathbf{T}'$ (cf. Section 1.3), will not contain any translations of formulas of $\mathbf{T}''$ unless these formulas are already in Int (F_2), the latter taken relative to $\mathbf{T}''$ [52].

This can be put still more plainly. Let us say that two formulas F_2 and G_2 of $\mathbf{T}''$ are *intensionally related* if

$$\vdash_{T''} (F_2 \to G_2) \text{ or } \vdash_{T''} (G_2 \to F_2),$$

and *intensionally unrelated* otherwise; and let us call them *intensional synonyms* if

$$\vdash_{T''} (F_2 \to G_2) \text{ and } \vdash_{T''} (G_2 \to F_2),$$

and *intensional heteronyms* otherwise. Then a faithful translation of $\mathbf{T}''$ in $\mathbf{T}'$ respects these four intensional categories: it maps inten-

---

51 In Section 2.3 we defined a faithful translation to be basically a syntactical interpretation enjoying a special proof-preserving property. We could as well define faithful translation basically in terms of translation in WANG's sense, the difference lying in the recursivity condition imposed on the inter-theoretic correspondence in the case of translation. The distinction is theoretically relevant, and our terminology — "faithful translation" instead of "faithful interpretation" — was chosen as a reminder of the possibility of thus sharpening the notion.

52 We of course understand that the rank of a formula is respected under interpretation or translation; the notion of rank is used in our definition of intension.

sionally unrelated formulas only onto intensionally unrelated formulas, heteronyms only onto heteronyms, and so on. Syntactical interpretations and simple translations do not respect all of these meaning categories.

This does not mean that a faithful translation of T'' in T' is an injection, for two intensionally synonymous formulas of T'' may well be mapped onto the same formula of T'; nor is it necessarily a surjection. But a faithful translation does induce an injection of intensions. Put even more concisely: while interpretation or simple translation only induce a *homomorphism* of intensions, faithful translation induces a *monomorphism* of the same.

Between formal theories, then, the notion of faithful translation seems to constitute a superior explicatum for the notion of meaning-preserving inter-theoretic correspondence, inasmuch as our explication of the intensional or linguistic component of meaning in formal theories is satisfactory. The sharpness of this result allowed us to indicate, in Section 2.3, how the relativity of our explanation of meaning of constructs in natural mathematical theories, obtained through formalization, can be softened: two distinct but equally valid formalizations of the same informally given mathematical structure should prove mutually faithfully translatable one in the other.

It is to be expected that an example of this would be given by the VON NEUMANN and the ZERMELO "reductions" of numbers to sets if carried out in formalized set theories. If both of these reductions capture the "same" structure, then the VON NEUMANN reduction should allow us to prove about numbers exactly what the ZERMELO reduction proves, no more and no less. Such a result would enhance the "objectivity" of the mathematical structure being formalized. Correlatively, a proliferation of distinct formal set theories can detract from the objectivity of the set concept.

But while the sharpness of the notion of faithful translation and of our explication of intension lets us know exactly what to look for in the way of meaning-preserving correspondences between formal mathematical theories, the question becomes progressively less clear as one gets farther away from the formal sphere.

In particular, it is imperative that we refine our analysis of the process of formalization given in Section 2.3, described there as the establishment of an informal faithful translation between a natural theory T' and a formal theory T. Unless the natural theory T' is singularly ripe for formalization, the epithet "informal" in this description must be interpreted very broadly, the essence of a successful formalization being that one can prove concerning the formalizations of the informal constructs exactly as much as one can prove concerning these informal constructs; that certain syntactical connectives and constructions be respected is secondary. Preservation of syntactical structure gains in importance as the crystallization in syntax of the intensional component of meaning in the natural theory progresses.

In this connexion, we should add that in a formalization situation, the appropriate correspondence effects an informal *isomorphism* of the intensional structure of the natural theory T' in the proof-theoretic structure of the formal theory T, since there is no reason for the formalization to contain formal constructs or relations of entailment which are superfluous relative to the natural theory. Furthermore, the glib assertion that a formalization is better protected from inconsistency than are natural theories is unacceptable. If the formalization is sufficiently faithful, it will be every bit as prone to contradiction as is the theory formalized — the inconsistency of FREGE's system is a case in point[53]. This may be forgotten when one looks only at the finished product, the formalization, but hopefully there will always be a KREISEL to call irresponsible formalists back to reality.

The imperfect compatibility of the formal and natural mathematical spheres can be very vividly pictured in terms of differential geometry, as THOM (1970) has done. He views a formalization T as having a certain zone of contact with the intuitive theory T' formalized: at the periphery of this contact zone, e. g., where the formulas of T become too long or too complex, making impossible their intuitive interpretation in T', there is a semantical take-off (*décollage sémantique*) which prohibits the extension beyond the contact zone of the informal meaning isomorphism which holds there between T' and T. The higher transfinite numbers seem to provide another example of a peripheral semantical take-off area.

THOM concludes that it is no more cause for surprise that some natural mathematical theories appear to be only locally axiomatizable and formalizable, i. e., that a particular formalization of such a theory may cover only a restricted semantical area of the theory,

---

53 An especially readable presentation of the formal aspect of FREGE's foundation is given by HATCHER (1968).

than it is disturbing to find that some differential manifolds cannot be covered by a single chart, i. e., that a single flat map cannot be made of some geometrical surfaces, such as the sphere, without "tearing", or that the earth is round.

This differential-geometric vocabulary seems well suited to bringing such semantical points home, and may be useful in dissipating somewhat the vagueness surrounding the notion of meaning-preserving correspondence in the context of natural mathematical theories. Indeed, the view of the proper role of a formalization hinted at in Section 2.3, as an *explanans* for the theory formalized (rather than as a foundation), becomes even more appealing here: in deciding whether or not two natural theories (geometrical surfaces) can be locally or globally mapped one onto the other by a meaning-preserving correspondence (without tearing), it can be of precious aid to first map through formalization (differential homeomorphisms) the fragments of the theories (surfaces) concerned onto a common formal (flat) ground to facilitate their comparison.

The flattening — the *ex-plan*-ation — of the theories thus obtained clarifies the constructs involved, and provides a unified view which may prove sufficiently suggestive to decide the questions of meaning equivalence: questions which it might, in certain cases, be virtually impossible to answer satisfactorily without formalization. It is in this perspective that we saw fit in Chapter 2 to promote the construal of meaning in a natural theory through the formal intensional component of meaning in a formalization of the theory. Faithful translation therefore appears to be an invaluable tool in explicating the notion of meaning-preserving correspondence between two theories, whether both are formal, or one is formal and the other natural, or even when both are natural.

Extensions are not taken into account in a faithful translation. Because of the secondary role which formal extensions play in determining meaning in mathematics, we do not consider their preservation to be a central issue in the discussion of meaning-preserving correspondences. This would not be so in a factual science, were a dualistic explication of meaning in terms of reference and entailment applied there. We did, however, allow in Section 2.4 that set-theoretic model theory may be utilized, not in explicating mathematical reference, but in giving a follow-up explication of formal mathematical meaning in a broadened semantical context. Let us briefly take up this theme.

Let $\tau\colon \mathbf{T}'' \to \mathbf{T}'$ be a syntactical interpretation or translation of the formal theory $\mathbf{T}''$ in the formal theory $\mathbf{T}'$. Let $\mathcal{M}''$ and $\mathcal{M}'$ designate the classes of set-theoretic models of $\mathbf{T}''$ and $\mathbf{T}'$ respectively. Any given model M' in $\mathcal{M}'$ induces through τ a model in $\mathcal{M}''$, which we designate as $\tau^*(M')$: the set-theoretic interpretation in $\tau^*(M')$ of the letters of $\mathbf{T}''$ is obtained by the composition of their syntactical interpretation (translation) in $\mathbf{T}'$ through τ, followed by the semantical interpretation of the resulting formulas (of $\mathbf{T}'$) in M'. Using our usual designations for reference relations, the extensions in $\tau^*(M')$ of formulas of $\mathbf{T}''$ are obtained through the composition of τ with ϱ_M, that is, the diagram

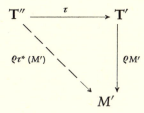

commutes.

There is here no guarantee, however, that all models of $\mathbf{T}''$ can be recognized as arising in this way from models of $\mathbf{T}'$. It is natural to consider syntactical interpretations or translations which do carry this guarantee of conservation of model-theoretic meaning: SVENO-NIUS (1972) calls such inter-theoretic correspondences *formally correct translations*.

Clearly the next questions to ask concern the relationship between this notion of correct translation, which is, so to speak, strictly conservative with respect to model-theoretic meanings, and that of faithful translation, which is strictly conservative with respect to proof-theoretic meanings. But for our present purposes we will content ourselves with noting that the notion of correct translation, relativized to restricted subclasses of $\mathcal{M}''$ and $\mathcal{M}'$, would seem to provide an interesting complement to faithful translation in explicating the notion of meaning-preserving inter-theoretic correspondence; much as, in Section 2.4, set-theoretic (or semantical) interpretation, taken relative to a class of "standard" models or to a particular "intended" model, may serve to complement the proof-theoretic explication of meaning obtained through the syntactical relation of entailment[54].

Thus we allow again that the explication of the intensional component of mathematical meaning need not be unidimensional. But we do feel that in discussing meaning-preserving correspondences, proof-theoretic considerations deserve precedence. The popularity of model-theoretic explications is perhaps due to a misconception of what constitutes objectivity in mathematics. We have argued that this objectivity resides more in proof-theoretic structure than in the notion of set.

4.3 Explanation v. Reduction

BENACERRAF has written recently that "the mathematician's interest stops at the level of structure. If one theory can be modeled in another (that is, reduced to another) then further questions about whether the individuals of one theory are really those of the second just do not arise ... the philosopher is not satisfied with this limited view of things. He wants to know more and does ask the questions in which the mathematician professes no interest ... And mistakenly so" (1965, p. 69). BENACERRAF's intent here is to rebuff the tendency to ascribe ontological import to the several reductions of the integers to set theory. Such statements are acceptable to this end, but in a larger context appear uncomfortably narrow.

For one thing, interest in and explication of what constitutes a reduction is itself a primary mathematical concern (BENACERRAF takes the notion for granted), albeit it takes a mathematician with a definite philosophical bent to find interest in carrying out a rigorous analysis of this supremely important informal concept[55]. For another, a mathematician's motivation for formulating new conjectures about the mathematical "structures" which figure in a given

---

54 For some material on the relation between faithful and correct translations, cf. HANF (1965), pp. 134 ff.

55 KREISEL's interest in reduction stems from his attempt to conceive a revised HILBERT programme, following which it would be possible to determine the "recursive content" or the "constructive equivalent" of classical mathematical concepts and theorems, not by means of the vague notion of constructive proof essential to HILBERT's formulation, but by using the notion of a reduction to a constructive formula, that is, a quantifier-free formula with only recursive non-logical constants.

reduction situation may be far removed from the level of theoretical, or deductive, structure; he may well find inspiration in the nebulously ontology-laden heuristic components of the theories involved. Let us therefore expand on these questions of reduction and structure with which the mathematician is constantly concerned.

We have suggested in the previous two sections how the informal notions of reduction and of meaning-preserving correspondence may be understood precisely, by giving formal explications of them. There remain many inter-theoretic correspondences in mathematics which are neither reductions nor faithful translations in this formal sense, e. g., strictly speaking, the notion of formalization, by its very nature, can only be described metaphorically as an informal faithful translation. It seems fitting to call such correspondences *explanations,* in that in explaining the meaning of a construct we may consider it permissible to refer to liberal and even extravagantly inexact interpretations, comparisons, and analogies, i. e., to what we called in Section 2.3 the quasi-translations which make up the heuristic component of a mathematical theory.

The explanation of one mathematical theory by another can be more or less precise, depending, to begin with, on the level of formal sophistication of these theories. It would be interesting to determine which "reductions" in mathematics can be given adequate formal expression as reductions or meaning-preserving correspondences in our sense: the process of refining an axiomatization, where it is recognized that certain primitive constructs or axioms may be dispensed with, could no doubt generally be recognized as a true reduction; the "associated statement form" reduction of the predicate calculus to the propositional calculus is clearly not a true reduction; HERBRAND's Theorem expresses a true reduction (in KREISEL's generalized sense); a formal theory expressed in classical predicate logic can be faithfully translated into the same theory expressed in a natural deduction logic; and so on. The "reductions" of various mathematical theories to set theory can, most likely, also be expressed formally as faithful translations.

The point here is that structure, and structure-preserving correspondences, are not always automatic mathematical "givens", except at the strictly formal level, where one has objective proof-theoretic structure and knows what must be done to preserve it. There may be a great deal of Kreiselian informal analysis to be done before a proposed informal "reduction" can be validated as such (or, more usually, as a meaning-preserving translation). The meanings of natural mathematical constructs are often vague, and may be built on ontological references (e. g., the infinitesimal, or concrete "sets"), and only gradually crystallize into objective relations of entailment through a series of reflective abstractions, or what could be called progressive "mathematizations".

In seeking to ascertain whether "structure" is preserved — in striving, first, to recognize structure — the mathematician does grapple with questions about whether the "individuals" (constructs) of one theory are really those of the second. Even if the reduction under examination is between formalized theories, the question whether the formalizations adequately capture the underlying natural theories must anteriorly have been resolved. Mathematical structure is not just there for the taking[56].

Because the development, or gradual objectivization, of mathematical structure is not easy, different explanations of a certain structure, gained through classical "reductions" or other less precise correspondences, are of great heuristic interest. It is invaluable to have several vantage points from which to develop the properties of the same structure. Considerable insight is afforded by such "representation theorems" as STONE's Theorem, relating algebraic and topological aspects of Boolean algebras. Or again, much is gained through the comparison of formal and natural treatments of the same structure, not only a clearer understanding of the natural notions, but also a sobering awareness of certain limitations of formalization; the benefits here are mutual.

---

56 Hence our preference, despite its realist overtones, for speaking of "natural" mathematical theories rather than "informal" ones. The latter adjective may leave the impression that there is just a simple easy step to be taken from a natural theory to its formalization.

A striking example of the force of a heuristic notion in propulsing a theory's development. while opposing at the same time a high degree of resistance to theoretical crystallization, is the contribution of the notion of the infinitesimal to mathematical analysis. After two centuries of empire the infinitesimal was demoted to the role of a cheap pedagogical device, only to be recently resurrected in all its glory by ROBINSON (1966a). This *dénouement* also points up the essential give-and-take which goes on between the theoretic and the heuristic: the heuristic serves to understand, develop and apply a theory, but can also receive a boost or promotion to higher respectability from theory in return.

Continuing a little further in this line of thought, one soon becomes aware that the mathematician is, in practice, at least as concerned with the *comparison* of structure as with its preservation. From the comparison of more or less different structures, from the exploration of their points of resemblance or dissimilarity, or again from the contrast between different stylistic views (e. g., set-theoretic and categorial) of the "same" structure, the mathematician derives clues as to which direction he should take in further pressing his investigations or constructions. Much more than the conservative desideratum of preservation of structure, or of a variable "naturalness", *suggestiveness* is the heart of the matter here.

In general we could cite each of the various situations which have been called "reductions" in mathematics as examples of the *cross-fertilization* of theories, where the theories involved take on added vigour and receive new development. Sometimes the gain is not immediately signalled by the appearance of new results, but there is almost always an immediate quantum jump heuristically, which in the long run may provide the insight needed to solve long-standing problems, e. g., the FERMAT-DESCARTES reduction of geometry to algebra proved essential to solving the problem of the squaring of the circle.

Thus while it would be expected that in a true reduction-situation, the reduced theory would be discarded and disappear in favour of the reducing theory, the opposite is more commonly the case. Both theories rather profit from the relation established between them, and may even spawn new hybrid theories, e. g. algebraic geometry. The logicist attempt to reduce mathematics to logic ultimately gave rise, on the one hand, to the expected increased utilization of logic in the organization and development of mathematical theories, but, on the other hand, resulted (rather ingloriously for the logicist doctrine) in the increasing mathematization of logic itself. The give-and-take is particularly fruitful here between logic and algebra.

Instead of true reductions, then, classical "reductions" are more frequently meaning-situations, or explanations, as GÖDEL also sees the process of formalization, in that insight into one mathematical structure is often gained, in these cases, by pointing to something else, to another structure. And the mathematician may even be legitimately interested in whether the "objects" of one theory are really those of the other, as, for example, in the case of the reduction of set theory to category theory (recall Section 3.7). Had the latter result chronologically preceded the intrinsic independence results in set theory, it would have contributed more importantly towards the de-ontologization of set theory.

Indeed, the reversal in attitude of set theorists with respect to the conception of truth, or "objectivity", in their discipline, from a quasi-empiricist correspondence view to a more conventionalist coherence view (cf. PUTNAM, 1968), cannot be dismissed as being of interest only to philosophers, and only to misguided ones at that, as BENACERRAF would have it. For this change in heuristic attitude opens the door to — in fact, encourages — the development of distinct set theories.

Each "live" mathematical theory is not only a deductive structure. Aside from the more or less formal deductive organization of its constructs, it comes equipped with its own particular heuristic charisma. While the exact notions of reduction and of faithful translation, as explicated above, provide efficient tools for comparing the meaning relations between constructs which have crystallized into relations of entailment in different theories, there can be no such effective guarantee of the transfer of the heuristic component of one theory to that of another.

Even under a true reduction, heuristic surprises can ensue. For example, a gain in "simplicity" (a very heuristic concept) may be achieved. Or two contradictory heuristic components may even be involved, as is the case in HERBRAND's Theorem, where one strongly feels, in virtue of the decidability of the reducing theory (propositional logic), that one has somehow found an Achilles' heel in the undecidability of the reduced theory — despite formal evidence to the contrary. These conflicting intuitions are eventually reconciled by setting up improved searching techniques for proofs in first-order logic, based on the HERBRAND result.

In any type of inter-theoretic correspondence, then, one can expect a certain amount of clash between the heuristic components involved. Far from being fatal to the correspondence in question, these collisions are most often occasions of rejuvenated insight, of interfecundation of the heuristic components in play. We might say that here truth thrives on contradiction, but on structured contradiction, in that the syntactic or linguistic components of meanings must be in reasonable accord under the given correspondence. But there is no way of "reducing" one theory's heuristics to another's, as heuristics follow no general pattern. Those unstructured but vital appendages of mathematical theories, namely, their heuristic components, escape such objective treatment, yet are no less important to mathematicians in so-called "reduction" situations.

Indeed, the whole interest of most of these "reductions" resides in the clash between the heterogeneous heuristic components involved, e. g., the reduction of geometry to algebra, of mathematics to set theory, of sets to categories or to numbers (the LÖWENHEIM-SKOLEM Theorem), of mathematics to "logic", and so on. Again, we lean here towards an Aristotelian view, as argued in Chapter 3.

What we have been saying amounts to declaring our preference for a kind of Principle of Specificity for mathematical theories, perhaps in an over-reaction to the currently popular sort of mathematical structuralism which fails to do justice to the (constructive) epistemological aspect of mathematics. Having criticized NAGEL's version of logical reduction as superficial, and having suggested a stronger explication in its place, we are still not content, and find that even inter-theoretic correspondences which can be recognized as proof-theoretic reductions, or as faithful translations, or even as correct translations or isomorphisms in HANF's sense (1965), may yet not be proper reductions, nor even, in the last analysis, proper meaning-preserving correspondences, at all, in as much as the *significations* (Section 3.6) of the concepts in play remain specific.

This principle of specificity of the heuristic component of a theory, partially manifested, perhaps, through a particular mathematical style, vaguely echos GLYMOUR's finding, in the physical sciences, that most cases of "reduction" relate together theories which are *in fact* mutually irreducible, and that we have not to do in these cases with true reductions or translations at all, but rather with instances of inter-theoretic explanation.

Though commonly classified as a tight-lipped formalist, BOUR-BAKI, at least once in his career, did not propose that we forget about "the meaning of the notions", as KREISEL would put it: "the mathematician does not work mechanically, like a worker on an assembly line; one cannot insist enough on the fundamental role played, in a mathematician's research, by a particular kind of *intuition,* not a common sensory intuition, but rather a kind of direct divination (preceding all reasoning) of the normal behaviour which he seems justified to expect of mathematical entities with which he has, through long experience, become almost as familiar as with the beings of the real world. Each structure brings with it its own proper language, all laden with special intuitive resonances . . ." (1962, p. 45).

That proper language is the one which the (creative) mathematician must gradually learn to recognize and to speak, and in terms of which he may ask questions concerning the "objects" (constructs) which inspire it. Interested in structure, the mathematician must first become conversant with its construction.

This proper language, this specificity of a mathematical theory, gives rise to a critical attitude towards reductionism in mathematics in general. In particular, it leads us to side with THOM, in speculating that "it is not sure that even in pure mathematics, every deduction may admit a set-theoretic model . . . Perhaps, even in mathematics, quality subsists, and resists all attempts at set-theoretic reduction. The old Bourbakist hope[57], to see the mathematical structures naturally come out of the hierarchy of sets, their subsets, and of their combinatorics, is, no doubt, only a chimera" (1970, p. 236). Recall THOM's view that different mathematical theories usually only admit *tangential* points of meaning identification, or *local* zones of semantic contact.

KREISEL has expressed a similar view: "Tame mathematicians [and tame philosophers!] have a rather soft conception of mathematical significance: if they know how to fit the intuitive [mathematical] notion into an existing frame work such as set theory, they call the notion significant and not otherwise" (1967b, p. 214). He goes on to make the interesting remark that it would be, in fact, very surprising were the continuum hypothesis solved by working only in the language of set theory. Since the problem concerns all possible subsets of the integers, why should one expect that the relevant properties be expressible set-theoretically?

Combining our view of what is and what is not preserved by an inter-theoretic explanation, or even by a true reduction or faithful translation in mathematics, with our conclusion from Chapter 3 that the only proper domain in which to conceive of the existence of mathematical objects is in the heuristics of mathematical theories, it is to be expected that we be sceptical of any talk of "ontological reduction" in mathematics. Since QUINE opines that mathematics is "up to its neck in commitments to an ontology of abstract enti-

---

57 Compare the previous quotation!

ties" (1953, p. 13), we will examine his ontological doctrines as he applies them to mathematics. But we remark first on a point related to our discussion of meaning reductions.

QUINE holds a dark view of the possibility of perfectly translating one language into another, and concludes with his familiar positions on the impossibility of radical translation and on referential inscrutability (cf. 1958 or 1960). Unless one is deluding oneself over one's powers to fathom meaning, there is no more reason for disappointment in these doctrines than there is in GÖDEL's Incompleteness Theorem or in CHURCH's Theorem. It is as foolish to expect the field linguist to seize at once the full meaning (extensional component included) of "gavagai", as it is natural to expect that through the sustained contact between natural languages, a phenomenon of cross-fertilization will lead to improved meaning correspondences between them, so that distinctions between "rabbit", "rabbithood" and "undetached rabbit part" which were perhaps possible only in one language become common and translatable in both — if the distinctions are vital enough.

Hope for eventually perfect translation and referential agreement springs in this way from the dynamic interaction of living languages, just as an imperfect initial "reduction" of one living mathematical theory to another may lead to the subsequent parallel development and more perfect harmonization of these theories. There is therefore no need to overemphasize the initial problems in translation: patience and ingenuity bring their fruit in the long run, just as they often turn to profit paradoxical conflicts in the heuristic components of theories in a reduction-situation.

4.4 Ontological Commitment

Any discussion of ontological inter-theoretic reduction must be preceded by a specification of definite ontologies for the theories involved. Indeed, for a reduction to be ontological, QUINE requires that it "preserve relevant structure" (1966b, p. 201). More precisely: "The standard of [ontological] reduction of a theory T'' to a theory T' can ... be put as follows. We specify a function, not necessarily in the notation of T'' or T', which admits as arguments all objects in the universe of T'' and takes values in the universe of T'. This is the *proxy function*. Then to each n-place primitive predicate of T'', for each n, we effectively associate an open sentence of T' in n free variables, in such a way that the predicate is fulfilled by a n-tuple of arguments of the proxy function always and only when the open sentence is fulfilled by the corresponding n-tuple of values" (ibid., p. 205, our italics) [58].

Application of this notion to mathematics presupposes that some general explication of mathematical reference has been given, in terms of which we may understand what QUINE means by "object" and "universe" of a mathematical theory. QUINE certainly makes no bones about attaching an ontological component to mathematical theories. Starting from the position that within natural science there is a "continuum" of gradations, from statements reporting observations to basic theoretic statements, e. g., of quantum theory, he extrapolates at both ends and holds that "statements of ontology or even of mathematics and logic form a continuation of this continuum ... The differences here are in my view differences only in degree and not in kind" (1951, p. 134). Since QUINE's doctrine of ontological commitment is the only general explication which he has offered to date of what could be his understanding of mathematical reference, we must first examine and evaluate this doctrine.

Concerning ontological commitment, "entities of a given sort are assumed by a theory if and only if some of them must be counted among the values of the variables in order that the statements affirmed in the theory be true" (1953, p. 103); or again, "what objects does a theory require? ... those objects that have to be values of variables for the theory to be true" (1969c, p. 96). Thus the *required objects* of a theory are those "values of variables" which must be available for the verification of the assertions of the theory, and the *universe* of a theory is made up of such required objects.

"Variables" enter the discussion because "the question of the ontological commitment of a theory does not properly arise except as that theory is expressed in classical quantificational form, or in so far as one has in mind how to *translate* it into that form. I hold this for the simple reason that the existential quantifier, in the objectual [referential] sense, is given precisely the existential interpretation and no other: there are things which are thus and so"

---

58 Clearly, to call such an inter-theoretic correspondence a "reduction" is a misnomer, in the strict sense of "reduction" (Section 4.1): but then, "ontological translation" is less catchy.

(1969c, p. 106, our italics). Or again, "referential quantification is the key idiom of ontology" (1969b, p. 66).

A first observation to be made here is that, in QUINE's view, the "reduction" of all mathematics to set theory, and the set-theoretic "semantics" of formalized theories, are to be swallowed hook, line, and sinker. This already sticks in our throat, since, for one thing, for reasons given above we question the validity of the transfer of the ontological (or semantical) commitment of a natural mathematical theory either laterally, to that of other natural theories, or vertically, to that of more formal theories, the heuristic component of a theory being to a greater or lesser degree specific to that theory.

For another thing, QUINE argues for the existential, or objectual, interpretation of the quantifiers against the substitutional interpretation, for purposes of construing the ontology of a theory (cf. 1969b, pp. 63 ff., and 1969c, pp. 104 ff.). On the substitutional interpretation, "$(\exists x) F(x)$" asserts that there is some closed term t in the formal language such that $F(t)$ holds.

Without closely examining QUINE's arguments here, we point out that PARSONS (1971a) has strongly challenged the exclusivity of the existential interpretation for ontological purposes, arguing that the substitutional interpretation does express a genuine conception of existence, and one which in fact appears especially appropriate to the manner of existence of "linguistic" abstract entities, such as classes, i. e., the extensions of predicates. PARSONS finds that in so far as one does not construe the classes involved as "real" independently of the expressions for them, the substitutional interpretation renders this concept of being perfectly well, and some more absolute verification or implication of their existence is senseless. This is indeed MARCUS' (1962) original point.

In yet another line of thought, taking up GÖDEL's distinction between physical and "abstract" objects, PARSONS (1971b) develops a constructivist mathematical ontology which illustrates a concept of existence different from QUINE's, which he qualifies as too "realistic". He advances that a distinction should be made between perceptible (physical) objects and abstract entities, or *Denkgebilde*, and that the best approach to the "question of the meaning of being" in the most elementary parts of mathematics is not through set theory: "Just by its universality, set theory leads us to neglect essential distinctions" (1971b, p. 152).

Taking as example a non-predicative reduction of predicative mathematics to natural numbers, PARSONS then shows how the introduction of abstract concepts, that is, concepts which apply to "objects" of a nature different from that of spatio-temporal forms, can be construed as an expansion of "ideology" beyond the intuitive-combinatorial without an expansion of ontology, i. e., without requiring such *Denkgebilde,* or "intensional" objects. PARSONS concludes that what a theory "says there is" is only a minimal characterization of what someone commits oneself to in accepting the theory.

But over and above these direct questionings of QUINE's doctrine of ontological commitment, which QUINE would tend to dismiss expeditiously as proposing "deviant" concepts of being, a second observation to be made, and which concerns us here, is that a sort of ontological reduction already holds a preponderant place in QUINE's presentation of his notion of commitment. Though ontology and mathematical logic must lie at opposite extremities of QUINE's continuous spectrum [59] of the natural sciences (cf. the above quotation from 1951), questions of ontological commitment are supposed to be docilely "translatable" all the way into predicate logic: "basic controversy over ontology can be translated upward into a semantical controversy about words and what to do with them" (1953, p. 16).

Such a view can be rationally held, and sense can be made of QUINE's definition of ontological commitment, only to the extent that a clear explication of such "translation" is given: which is to say, or so it seems to us, only inasmuch as a certain notion of ontological reduction is already clear.

In point of fact, QUINE convinces us that questions of ontology are understood in terms of background theories by "paraphrase into some antecedently familiar vocabulary" (1969b, p. 54), and that "it makes no sense to say what the objects of a theory are, beyond saying how to interpret or reinterpret that theory in another" (ibid., p. 50). Put most pertinently: "We are finding no clear difference be-

---

59 PIAGET's findings, that the objectivity in mathematics arises from the invariance of elementary operations on objects rather than from the objects themselves (1970b, pp. 18 ff.), would seem to seriously throw into question this continuity, in differentiating anontological mathematics and logic from the factual natural sciences.

tween specifying a universe of discourse — the range of the variables of quantification — and *reducing* that universe to some other" (ibid., p. 43, QUINE's italics).

In seeking to understand ontological commitment we are therefore caught in a squirrel-cage of "translating upwards", "translating downwards" (into a background theory), "interpreting", "paraphrasing", "reducing", and what have you. Rather than treat the problem squarely, QUINE is somewhat flippant about the notion of ontological paraphrase, letting an air of subjectivity hover about it, referring to a paraphraser as playing the game, and intimating that someone who refuses to paraphrase is a sore loser (1960, p. 243). The informal atmosphere that is a large part of QUINE's style masks the fact that some sort of ontological reduction must apparently already be accepted before ontological commitment is.

And even supposing that after running the gauntlet of paraphrase, one somehow arrives at the level of quantification theory with some shreds of ontology still in one's grasp, one's problems are not yet solved, as QUINE is even less specific about what it means to be a "value of a variable", and seems to simply take for granted that there is some theory of reference behind this phrase. One can understand PRIOR's asserting that "'to be a value of a bound variable is to be' is just a piece of unsupported dogma" (1971, p. 48).

Of course, QUINE has in mind set-theoretic "semantics" as providing an ontological hinterland for quantification theory. But aside from our many reasons laid out above for refusing the validity of such a move, even if we decide to play the game with QUINE, he is not clear on whether values are to be assigned to variables following a fixed reference relation ϱ_M attached to a particular model M, or following a relation $\varrho_{\mathcal{M}}$. In one pertinent passage, QUINE asserts: "there is more to be said of a theory, ontologically, than just saying what objects, if any, the theory requires; we can also ask what various universes would be severally sufficient. The specific objects required, if any, are the objects common to all those universes" (1969c, p. 96). But then, in terms of set-theoretic semantics, *no* objects would ever be "required" by a theory, for one can trivially construe models for the same theory which are isomorphic but which have disjoint universes.

One can try to resolve this paradox by reading further: "of course a theory may ... require no objects in particular, and still not tolerate an empty universe of discourse either, for the theory might be fulfilled equally well by either of two mutually exclusive universes. If for example the theory implies "($\exists x$) (x is a dog)", it will not tolerate an empty universe; still the theory might be fulfilled by a universe that contained collies to the exclusion of spaniels, and also *vice versa*" (ibid., p. 96).

But to thus say that a theory may require "no objects in particular" makes one wonder what sense there is in speaking here of ontology at all. QUINE's own words, used in shooting down substitutional quantification as rendering a defendable concept of being, would seem to bear witness against his own notion of ontological commitment: "Substitutional quantification makes good sense ... no matter what substitution class we take ... To conclude that entities are being assumed that trivially ... is simply to drop ontological questions" (1969c, p. 106).

It seems to us that QUINE is dealing not with existing nor even with possible objects, but rather with kinds of objects and defining descriptions. In our opinion, the explication of such notions is most properly carried out entirely at the level of syntax (recall our discussion of RESCHER's explication of possibles in Section 2.7). Or again, if one insists on a model-theoretic approach, then KREISEL's work (1965) on explicating the notion of "the common part" (which he finds reason to call the *hard core*!) of models in a given class of models, and on relating this notion to definability theory, illustrates how this can be quite palatably done without smothering the issue in a doubtful ontological gravy[60].

---

60 Ontological commitment as an explication for reference has nevertheless inspired several philosophers. For example, CHENG has proposed the following programme: "Now we can show that the inscrutability of reference can be reconstrued as requiring that a given language (or theory), if being referential in admitting an ontology, cannot be formalized ... Analyzing formalizability, we can relate some metamathematical notions to the question of reference and use them as precise criteria and conditions for referentiality" (1969, p. 783). This is like using GÖDEL's incompleteness results to argue that mind is not mechanical without first elaborating a theory of mind.

In like vein, JUBIEN takes formal semantics as a theory of reference for formalized theories (both factual and nonfactual): "vivid examples are available if one turns to formal theories of *physical* objects, for here the notion that all models are on an equal footing is visibly absurd" (1969, p. 540). JUBIEN explicates the notion of ontological reduction by his

Whatever may be the case, it is evident that if QUINE's "value of a variable" is interpreted using the set-theoretic semantics of first-order theories, as QUINE certainly intends it to be, then no theory ever requires or is ontologically committed to more than a countable universe of objects, by the LÖWENHEIM-SKOLEM Theorem. Though QUINE refuses to accord this theorem any value as an ontological reduction, as it fails to specify a proxy function (1966b, p. 206), it nevertheless appears rigorously applicable to his notion of ontological commitment. Consequently it seems to follow that the "objects" which QUINE entertains a theory as referring to when discussing ontological reduction are not to be confused with the "required objects" to which the theory is ontologically committed — indeed, there are only a countable number of formulas of the form $(\exists x)\, F(x)$ in any first-order theory.

The overall conclusion which emerges from all this is that when discussing ontological reduction, QUINE has in mind some sort of specific ontologies, whereas ontological commitment is largely a matter of first-order syntax. This is unfortunate, as it would seem preferable that the explications of ontological commitment and of ontological reduction should dovetail and reinforce each other, e. g., as did our explications of meaning and of meaning reduction. But separation of the "paraphrase" which accompanies the determination of ontological commitment from the "reduction" in ontological reduction, and separation of the "objects" involved in ontological reduction from the "required objects" of commitment, cannot be avoided, as the two kinds of "reduction" and the two kinds of "objects" appear irreconcilable from QUINE's own treatment of them.

This frees us from the squirrel-cage, but puts the onus on QUINE to urgently provide *two* explications, namely, of the "objects" in terms of which ontological reduction is defined, and of the "paraphrase" by means of which ontological commitment is determined. Having recognized that there are two problems here rather than one, let us return to the former.

---

notion of a "conceptual reduction", which is nothing else than syntactical interpretation in our sense, and which, as we have seen, can neither be properly called a "reduction", nor is it "conceptual", in that it need not preserve intensions: and there is most certainly nothing ontological about it.

4.5 Ontological Reduction

We therefore abandon trying to construe the "objects" involved in ontological reduction in the "light" of ontological commitment, in favour of a pragmatic approach. This clears the air considerably, as scrutiny of the examples of ontological reduction which QUINE (1966b) gives — the FREGE and VON NEUMANN reductions of the natural numbers to sets, and various reductions of the reals such as DEDEKIND's — reveals that QUINE only conceives ontological reduction in terms of theories which already have an "ontology", and this in no exact sense but rather in a vague, heuristic sense.

For QUINE, however, such ontologies appear quite tangible. For though he admits that "Numbers . . . are known only by their laws, the laws of arithmetic . . . Sets in turn are known only by their laws, the laws of set theory" (1969b, p. 44), he nevertheless attaches somehow an objective ontological component to these theories: "what justifies an ontological reduction is, vaguely speaking, preservation of relevant structure. What we now perceive is that this relevant structure runs deep; the objects of the one system must be assigned severally to objects of the other" (1966b, p. 207). How we are to reconcile these two statements is, to say the least, perplexing.

We could say that, for QUINE, some mathematical structures consist of more than a deductively organized theory — though we can "know" (or "perceive"?) these structures only through their "laws". Some other mathematical structures probably do not extend, iceberg-like, beneath a deductive, or intensional, surface. For example, QUINE does not speak of ontologically reducing the theory of abelian groups to that of rings, though this would be a simple matter, if "objects" were "required objects"; for the only object required by abelian group theory is an identity element, and in recognizing the group axioms to be a subset of those for rings, one considers this one object as mapped (by a "proxy" function) onto the required additive identity of ring theory. Or even if "objects" were not to be so taken, we may systematically construe any abelian group as a ring by adding to it a trivial multiplicative structure.

But what to make, then, of QUINE's not mentioning such obvious reductions as that of arithmetic to, say, the set of numbers larger than 16, a reduction which is apparently as ontological as numbers can be? Indeed QUINE employs this example as such in a different context, in illustrating his point of "ontological relativity" (1969b, p. 54). As THARP (1971) has observed, QUINE's choice of examples in illustrating his notion of ontological reduction leads us to feel that something other than ontology is at stake in these examples, perhaps something as vague yet as significant as the notion of explanation (Section 4.3).

What is also perplexing, is that when a theory T'' is syntactically interpreted in another theory T', an "ontological reduction" is induced between the ontologies of these theories, but in the direction opposite to that of the syntactical interpretation: any model of T' can be construed as a model of T'' (cf. the diagram in Section 4.2). From this perspective, the ontological reduction accompanying a syntactical interpretation or translation proceeds in a reverse orientation. It can therefore be argued *secundum naturam,* as the categorists would put it, that ontological, or extensional, reductions follow a course *inverse* to that of ideological, or intensional, reductions.

Let us look closer at QUINE's description of the FREGE and DEDEKIND reductions: "Here the proxy function f is the function which, applied, e. g., to the "genuine" number 5, gives as value the class of all five-member classes ... In general $f(x)$ is describable as the class of all x-member classes. When the real numbers are reduced to classes of ratios, $f(x)$ is the class of all ratios less than the "genuine" real number x" (1966b, p. 206).

These same "reductions" can just as well be described as meaning-preserving inter-theoretic correspondences, without taking any deep ontological plunge. A theoretical mode of description seems even more indicated in the case of the arithmetization of syntax accomplished by Gödelization, which QUINE oddly considers to be an outstanding example of ontological reduction, "expressions or strings of signs" being mapped into "the universe of elementary number theory, which consists of numbers" (1969b, p. 56), with assignment of Gödel numbers playing the role of the proxy function. This reduction of syntax to arithmetic can well do without such excessive "marksist" ontologization.

We prefer to construe the classical "reductions" in mathematics as more or less perfect *meaning-preserving* inter-theoretic correspondences, i. e., syntactical interpretations, faithful translations, Kreiselian proof-theoretic reductions, correct translations or isomorphisms in the model-theoretic sense, and the like. Such notions accomplish all that QUINE's can handle. In addition they are more sensitive, for QUINE does not strongly distinguish between strict and broad senses of reduction. They are more efficient, as they do without ontology (in view of the independence results, just what sort of object is a "set"?). They are more general, as they make sense for any pair of theories, whereas QUINE's notion only applies where there is some established "ontology". They are more satisfactory as they fit well with a general explication of meaning (cf. Sections 2.3, 2.4, 4.2), and should for the same reason be more fruitful. And above all they are more precise.

One could hope that QUINE would be agreeable to this conclusion, especially since he expects a proxy function to make exactly the distinctions made by the theory being reduced, and no more (1969b, p. 56), and so seems open to an intensional treatment of reduction. After all, syntactical interpretations and translations map an existentially quantified formula onto a similar formula, and so preserve ontological commitment. Also, the case against over-interpreting the LÖWENHEIM-SKOLEM Theorem can be strengthened and simplified through the use of our notions.

QUINE's argument against interpreting the LÖWENHEIM-SKOLEM Theorem as an ontological reduction hinges on the fact that nowhere in the various proofs of that theorem is there anything recognizable as a proxy function: "I see in the proof even of the strong LÖWENHEIM-SKOLEM Theorem no reason to suppose that a proxy function can be formulated anywhere" (1969b, p. 60). This type of argument is not convincing. Surely the point is not that after careful search one does not come up with an explicit proxy function (how does one know, outside of a circular appeal to good sense, that someone will not come up with a "constructive" proof, exhibiting a proxy-like function?), but rather that there is no denumerable ontology here at all, not even in QUINE's sense, nor is there any ontology being reduced.

The theorem simply establishes that under certain conditions, a formal theory T'' can be semantically interpreted in a set-theoretic model with a countable universe U'. There is no reducing theory T' involved, of which U' would be the natural "ontology", and there is no question of a chosen "ontology" U'' of T'' being reduced. The LÖWENHEIM-SKOLEM Theorem can thus be easily and transparently disarmed of its supposed ontological implications, without resorting to QUINE's proxy functions. The situation is different with the LSGB Theorem (cf. Section 4.1), which can be construed as establishing a "reduction" between any first-order theory and a special kind of formal number theory. It would be interesting to know whether QUINE would consider the LSGB Theorem as establishing a Pythagorean reduction of mathematics to numbers. JUBIEN (1969) admits not knowing how to deal with this result in terms of Quinean notions.

Another sensitive point in QUINE's notion of reduction, and which he himself candidly points out (1966b, p. 206, and 1969b, p. 58), is that there is a definite circularity attached to ontological reduction, in that the existence of the entities to be reduced must first be assumed in order to carry out the reduction. His assurance that a *reductio ad absurdum* attitude is nevertheless as justifiable here as in other contexts of logical argument leaves something to be desired.

We touch here upon the fundamental difficulty of how to make sense of the concept of ontological reduction. Is the proxy function, like the magician's handkerchief, supposed to somehow whisk already existing objects out of existence? Recall that we are dealing, in the context of ontological reduction, with established ontologies, with "existing" objects, not with "required" objects. Consequently ontological reduction does *not* establish that "if all of U were needed then not all of U would be needed" (1969b, p. 58; U is the universe of the theory being reduced), but rather that "for the theory T'' with universe U'', there has been found a theory T' with universe U', a correspondence between the formulas of T'' and the formulas of T', and a correspondence between the objects of U'' and the objects of U', such that ... etc.".

This tells us that we can simulate perfectly, with the theory T' and its universe, everything that we can already say or do with the theory T'' and its objects. But this in no way forces the objects of U'' out of existence. Or, as WANG puts it: "There is no reason to suppose that numbers evaporate but classes are rocks" (1966, p. 341). In fact, in so far as our "explication" of mathematical existence in terms of heuristics is correct, it is rather the opposite which occurs: *both* ontologies take on new life, there is a mutual fecundation of heuristic components. WANG makes essentially the same point: "Even if, after a proof of a theorem in number theory has been discovered, it is possible to eliminate defined terms and translate the proof into the primitive notation of set theory, the translated proof would not have been discovered by one who worked exclusively with the primitive notation of set theory" (ibid.,

p. 339). While proof-theoretic meaning is translatable, heuristics — and ontology — are not.

Whereas this circularity attendant to the notion of reduction proves fatal in our eyes to the notion of ontological reduction, it is not so for KREISEL's explication of reduction. For KREISEL can stipulate (1964, p. 171) that the adequacy of the translation of T'' in T' in a true reduction must be established using only the proof-theoretic means of the reducing theory T'. Thus circularity can be attenuated in this case, and proof-theoretic meaning proves properly reducible, thanks to our having an objective understanding of the proof-theoretic power of a theory. The efficacy of the mere existence of a proxy function in bringing about a veritable ontological reduction is open to doubt for lack of a similarly independently explicated notion of ontology.

But no matter how well one argues the superior coherence of construing the classical reductions in mathematics as intensional reductions, or as explanations, instead of as ontological reductions, it is to be expected that QUINE will doggedly go on waving the magician's wand: "The definitions of numbers by FREGE and VON NEUMANN are best seen as ontological reductions" (1966b, p. 200). QUINE apparently wishes his doctrine of referential inscrutability, which has its points in factual contexts, to apply also somehow to mathematics, by insisting on the use of classical quantification theory as the "key idiom of ontology", and thus leaving room for vague ontological speculation on the basis of more powerful background theories. Without even needing to go so far as THOMASON (1971), who, carefully interpreting QUINE all the way, finds at the end that reference is nonsense, one feels that here Occam is wielding his razor in thin air.

QUINE fails to show how his referential way of speaking can be efficient or fruitful in mathematics, and has yet to clearly explicate his notion of the ontology of a mathematical theory. What makes QUINE's mathematical examples of "ontological reduction" significant, for us, are the interesting clashes and possible cross-fertilizations of the heuristic components involved, instead of some putative sense of ontological economy. THARP (1971) concludes similarly. Its wide gamut of applications notwithstanding, set theory is one mathematical theory among others, and indeed holds no monopoly on the sort of Platonist heuristic *milieux* which gives rise to ontological entertainment, and, more importantly, sometimes to theoretical advancement in mathematics.

In discussing the notions of intension, formalization, style, and reduction, we have described how a certain *semantic ascent* may be considered to be operative in mathematics, in the sense of an upwards translation of meanings, or again, as KREISEL (1967b) puts it, as a derivation of proof from meaning, rather than as a transmission of ontologies (compare QUINE, 1960, pp. 270 ff.). Our difference with QUINE on the matter of ontological reduction turns on his distrust of (intensional) meaning and enthusiasm for (extensional) reference. From Chapter 1 onwards, we have tried to show how intensions play in fact a primal role in the semantics of mathematics, and how set-theoretic models, though they do provide a satisfactory formal explication of extension, may more properly be taken as offering a higher-order complement to the determination of (intensional) meaning in mathematics.

All in all, we find an overnarrow identification of semantics with reference regrettable, and QUINE's ontological doctrines, based on the triple association of set theory, semantics, and reference, most artificial in a mathematical context. On this point we would be tempted to reverse QUINE's slogan, that in mathematics a "murky intensionality of attributes" gives way to a "limpid extensionality of classes" (1958, p. 21), to the slogan that, in mathematics, the limpid intensionality of predicates replaces the murky extensionality of sets. From a *logical* point of view, i. e., aside from heuristics, in mathematics there is no speaking ontologically, and no speaking of ontology.

Bibliography

ADDISON, J. W., HENKIN, L., and TARSKI, A., eds., 1965: The Theory of Models. Amsterdam: North-Holland.

ALSTON, W. P., 1958: Ontological Commitments. In: BENACERRAF and PUTNAM, 1964, pp. 249—257.

ARNAULD, A., and NICOLE, P., 1662: La logique ou l'art de penser. Stuttgart: Fromann, 1965.

BENACERRAF, P., 1965: What Numbers Could Not Be. Philosophical Rev. 74, 47—73.

— and PUTNAM, H., eds., 1964: Philosophy of Mathematics. Englewood Cliffs, N. J.: Prentice-Hall.

BERNAYS, P., 1935: On Platonism in Mathematics. In: BENACERRAF and PUTNAM, 1964, pp. 274—286.

— 1946: Quelques points de vue concernant le problème de l'évidence. Synthèse 5, 321—326.

— 1950: Mathematische Existenz und Widerspruchsfreiheit. In: Etudes de Philosophie des sciences en hommage à F. GONSETH, pp. 11—25. Neuchâtel: Griffon.

— 1967: What Do Some Recent Results in Set Theory Suggest? In: LAKATOS, 1967, pp. 109—112.

— 1970: Die schematische Korrespondenz und die idealisierten Strukturen. Dialectica 24, 53-66.

BETH, E. W., 1956: L'existence en mathématiques. Paris: Gauthier-Villars.

— 1962: Extension and Intension. In: Logic and Language, pp. 64—68. Dordrecht, Holland: D. Reidel.

— 1966: The Foundations of Mathematics. New York: Harper and Row.

— and Piaget, J., 1961: Epistémologie mathématique et psychologie. Paris: Presses Universitaires de France. English translation by W. MAYS, Mathematical Psychology and Epistemology. New York: Gordon and Breach, 1966.

BIRKHOFF, G., 1966: Lattice Theory. Providence, R. I.: American Mathematical Society.

BOOLOS, G., 1971: The Iterative Conception of Set. J. Philosophy *68*, 215—231.

BOURBAKI, N., 1962: L'architecture des mathématiques. In: F. LE LIONNAIS, ed., Les grands courants de la pensée mathématique, 2nd ed., pp. 35—47. Paris: Blanchard.

BUNGE, M., 1967: Scientific Research. New York: Springer-Verlag.

— 1972: A Program for the Semantics of Science. To appear in: J. Philosophical Logic.

CANTOR, G., 1883: Grundlagen einer allgemeinen Mannigfältigkeitslehre. In: E. ZERMELO, ed., Georg Cantor Gesammelte Abhandlungen, pp. 165—209. Hildesheim, West Germany: G. Olms.

CARNAP, R., 1937: The Logical Syntax of Language. New York: Harcourt Brace.

— 1939: Foundations of Logic and Mathematics. Chicago: University of Chicago Press.

— 1950: Empiricism, Semantics, and Ontology. In: BENACERRAF and PUTNAM, 1964, pp. 233—248.

— 1955: Meaning and Synonymy in Natural Languages. In: CARNAP, 1956, pp. 233—247.

— 1956: Meaning and Necessity, 2nd edition. Chicago: University of Chicago Press.

CASTONGUAY, C., 1972: Naturalism in Mathematics. To appear in: J. Philosophical Logic.

CHENG, C. Y., 1969: Referentiality and Its Conditions. Abstract in: J. Philosophy *66*, 783.

CHURCH, A., 1944: Review of Lewis (1944). J. Symbolic Logic *9*, 28—29.

COHEN, P. J., 1966: Set Theory and the Continuum Hypothesis. New York: Benjamin.

— 1971: Comments on the Foundations of Set Theory. In: SCOTT, 1971, pp. 9—15.

CURRY, H. B., and FEYS, R., 1958: Combinatory Logic. Amsterdam: North-Holland.

DAVIDSON, D., 1967: Truth and Meaning. Synthèse *17*, 304—323.

EDWARDS, P., ed., 1967: The Encyclopedia of Philosophy, 6 vols. New York: MacMillan and Free Press.

FEFERMAN, S., 1969: Set-theoretical Foundations of Category Theory. In: S. MAC LANE, ed., Reports of the Midwest Category Seminar III, pp. 201—247. Berlin—New York: Springer-Verlag.

FEYERABEND, P., 1962: Explanation, Reduction, and Empiricism. In: H. FEIGL and W. SELLARS, eds., Minnesota Studies in the Philosophy of Science, Vol. 3, pp. 28—97. Minneapolis, Minn.: University of Minnesota Press.

FRAASSEN, B. VAN, 1967: Meaning Relations among Predicates. Noûs 1, 161—179.

FREYD, P., 1965: The Theories of Functors and Models. In: ADDISON, HENKIN, and TARSKI, 1965, pp. 107—120.

FREYTAG-LÖRINGHOFF, B. VON, 1951: Philosophical Problems of Mathematics. New York: Philosophical Library.

FRISCH, J. C., 1969: Extension and Comprehension in Logic. New York: Philosophical Library.

GLYMOUR, C., 1970: On Some Patterns of Reduction. Philosophy of Science 37, 340—353.

GOBLE, L. F., 1967: Abstract of A Coherence Theory of Meaning. Doctoral dissertation, University of Pittsburgh. Dissertation Abstracts 28, 5104-A.

GÖDEL, K., 1944: RUSSELL's Mathematical Logic. In: BENACERRAF and PUTNAM, 1964, pp. 211—232.

— 1964: What is Cantor's Continuum Problem? In: BENACERRAF and PUTNAM, 1964, pp. 258—273.

GOGUEN, J. A., 1969: The Logic of Inexact Concepts. Synthèse 19, 325—373.

— 1970: Mathematical Representation of Hierarchically Organized Systems. In: Global Systems Dynamics. New York: S. Karger.

GONSETH, F., 1936: La logique en tant que physique de l'objet quelconque. In: Actes du Congrès international de Philosophie scientifique, pp. 1—24. Paris: Hermann.

— 1948: Les conceptions mathématiques et le réel. Archives de l'Institut international des Sciences théoriques, série A, n. 2, 31—49.

— 1970: Mon itinéraire philosophique. Revue internationale de philosophie, nos. 93—94, fasc. 3—4.

GOODMAN, N., 1956: A World of Individuals. In: BENACERRAF and PUTNAM, 1964, pp. 197—210.

GOODSTEIN, R. L., 1968: Existence in Mathematics. Compositio Mathematica 20, 70—82.

— 1969: Empiricism in Mathematics. Dialectica 23, 50—57.

GRANGER, G.-G., 1968: Essai d'une philosophie du style. Paris: A. Colin.

GRASSMANN, H. G., 1844: Die lineare Ausdehnungslehre, ein neuer Zweig der Mathematik: In: Gunther Grassmann Gesammelte Werke. Leipzig: Teubner, 1894.

HANF, W., 1965: Model-theoretic Methods in the Study of Elementary Logic. In: ADDISON, HENKIN, and TARSKI, 1965, pp. 132—145.

HART, W. D., 1970: Skolem's Promises and Paradoxes. J. Philosophy 67, 98—109.

HATCHER, W. S., 1968: Foundations of Mathematics. Philadelphia: Saunders.

HEIJENOORT, J. VAN, 1967: Gödel's Theorem. In: EDWARDS, 1967.

HEMPEL, C., 1948: Problems and Changes in the Empiricist Criterion of Meaning. In: LINSKY, 1952, pp. 163—185.

— 1950: Review of Lewis (1946). J. Symbolic Logic 13, 40—45.

HENKIN, L., 1950: Completeness in the Theory of Types. In: HINTIKKA, 1969, pp. 51—63.

— 1955: The Representation Theorem for Cylindrical Algebras. In: Mathematical Interpretation of Formal Systems, pp. 85—97. Amsterdam: North-Holland.

—, MONK, D., and TARSKI, A., 1971: Cylindric Algebras. Amsterdam: North-Holland.

HINTIKKA, J., 1968: Logic and Philosophy. In: KLIBANSKY, 1968, pp. 3—30.

— 1969: The Philosophy of Mathematics. London: Oxford University Press.

JUBIEN, M., 1969: Two Kinds of Reduction. J. Philosophy 66, 533—541.

KALMÁR, L., 1967: Foundations of Mathematics — Whither Now? In: LAKATOS, 1967, pp. 187—207.

KAUPPI, R., 1967: Einführung in die Theorie der Begriffsysteme. Tampere: Tampereen Yliopisto.

KEYNES, J. N., 1887: Formal Logic, 2nd edition. London: MacMillan.

KLEENE, S. C., 1952: Introduction to Metamathematics. Princeton, N. J.: Van Nostrand.

KLIBANSKY, R., ed., 1968: Contemporary Philosophy, Vol. 1. Firenze: Nuova Italia Editrice.

KNEALE, W., and KNEALE, M., 1962: The Development of Logic. London: Oxford University Press.

KÖRNER, S., 1960: The Philosophy of Mathematics. New York: Harper and Row.

— 1967: Reply to Mr. Kumar. British J. Philosophy of Science 18, 323—324.

KREISEL, G., 1953: A Variant to Hilbert's Theory of the Foundations of Mathematics. British J. Philosophy of Science 4, 107—129.

— 1955: Models, Translations, and Interpretations. In: Mathematical Interpretation of Formal Systems, pp. 26—50. Amsterdam: North-Holland.

KREISEL, G., 1958: Wittgenstein's Remarks on the Foundations of Mathematics. British J. Philosophy of Science *9*, 135—158.

— 1964: Hilbert's Programme. In: BENACERRAF and PUTMAN, 1964, pp. 157—180.

— 1965: Model-Theoretic Invariants: Applications to Recursive and Hyperarithmetic Operations. In: ADDISON, HENKIN, and TARSKI, 1965, pp. 190—205.

— 1967a: Informal Rigour and Completeness Proofs. In: LAKATOS, 1967, pp. 138—186.

— 1967b: Mathematical Logic: What Has It Done for the Philosophy of Mathematics? In: R. SCHOENMAN, ed., Bertrand Russell: Philosopher of the Century, pp. 201—272. London: George Allen and Unwin.

— 1969: Two Notes on the Foundations of Set-Theory. Dialectica *23*, 93—114.

— 1970: The Formalist-Positivist Doctrine of Mathematical Precision in the Light of Experience. L'Age de la science *3*, 17—46.

— 1971: Observations on Popular Discussions of Foundations. In: SCOTT, 1971, pp. 189—198.

KRIPKE, S., 1965: Semantical Analysis of Intuitionistic Logic I. In: V. N. CROSSLEY and M. A. E. DUMMETT, eds., Formal Systems and Recursive Functions, pp. 92—130. Amsterdam: North-Holland.

KUMAR, D., 1967: Logic and Inexact Predicates. British J. Philosophy of Science *18*, 211—222.

LADRIÈRE, J., 1957: Les limitations internes des formalismes. Paris: Gauthier-Villars.

LAKATOS, I., 1962: Infinite Regress and the Foundations of Mathematics. Aristotelian Society Supplementary Vol. 36, 155—184.

— 1963: Proofs and Refutations (I)—(IV). British J. Philosophy of Science *14*, 1—25, 120—139, 221—245, 296—342.

—, ed., 1967: Problems in the Philosophy of Mathematics. Amsterdam: North-Holland.

LAWVERE, F. W., 1966: The Category of Categories as a Foundation of Mathematics. In: Proceedings of the Conference on Categorical Algebra, La Jolla, 1965, pp. 1—20. New York: Springer-Verlag.

— 1969: Adjointness in Foundations. Dialectica *23*, 281—296.

LEWIS, C. I., 1944: The Modes of Meaning. In: LINSKY, 1952, pp. 50—63.

— 1946: An Analysis of Knowledge and Valuation. La Salle, Ill.: Open Court.

— 1951: Notes on the Logic of Intension. In: P. HENLE et al., eds., Structure, Method, and Meaning: Essays in Honour of H. M. Sheffer, pp. 25—34. New York: Liberal Arts Press.

LINSKY, L., ed., 1952: Semantics and the Philosophy of Language. Urbana, Ill.: University of Illinois Press.

LYNDON, R., 1966: Notes on Logic. Princeton, N. J.: Van Nostrand.

MAC LANE, S., 1968: Category Theory. In: KLIBANSKY, 1968, pp. 286—293.

— 1971: Categorical Algebra and Set-theoretic Foundations. In: SCOTT, 1971, pp. 231—240.

— and BIRKHOFF, G., 1967: Algebra. New York: MacMillan.

MARCUS, R. BARCAN, 1962: Interpreting Quantification. Inquiry 5, 252—259.

MARTIN, R. M., 1964: On Connotation and Attribute. J. Philosophy 61, 711—724.

MATES, B., 1970: Review of White (1967). J. Symbolic Logic 35, 303.

MEHLBERG, H., 1960: The Present Situation in the Philosophy of Mathematics. Synthèse 12, 380—414.

MONTAGUE, R., 1968: Pragmatics. In: KLIBANSKY, 1968, pp. 102—122.

MOODY, E. W., 1953: Truth and Consequence in Mediaeval Logic. Amsterdam: North-Holland.

MOSS, J. M. B., 1971: Kreisel's Work on the Philosophy of Mathematics — I. Realism. In: R. GANDY and C. YATES, eds., Logic Colloquium '69, pp. 411—438. Amsterdam: North-Holland.

MOSTOWSKI, A., 1967: Recent Results of Set Theory. In: LAKATOS, 1967, pp. 82—96.

MYHILL, J., 1951: On the Ontological Significance of the Löwenheim-Skolem Theorem. In: I. M. COPI and J. A. GOULD, eds., Contemporary Readings in Logical Theory, pp. 40—51. New York: MacMillan.

NAGEL, E., 1944: Logic Without Ontology. In: BENACERRAF and PUTNAM, 1964, pp. 302—321.

— 1961: The Structure of Science. New York: Harcourt, Brace and World.

PARSONS, C., 1971a: A Plea for Substitutional Quantification. J. Philosophy 68, 231—237.

— 1971b: Ontology and Mathematics. Philosophical Rev. 80, 151—176.

PIAGET, J., 1970a: L'épistémologie génétique. Paris: Presses Universitaires de France.

— 1970b: Genetic Epistemology. New York: Columbia University Press.

— and INHELDER, B., 1969: The Gaps in Empiricism. In: A. KOESTLER and J. R. SMYTHIES, eds., Beyond Reductionism, pp. 118—160. London: Hutchinson.

POLYA, G., 1962: Mathematical Discovery. New York: Wiley.

POSZGAY, L., 1971: Liberal Intuitionism as a Basis for Set Theory. In: SCOTT, 1971, pp. 321—330.

PRIOR, A. N., 1971: Objects of Thought. London: Oxford University Press.

PUTNAM, H., 1968: Foundations of Set Theory. In: KLIBANSKY, 1968, pp. 275—285.

QUINE, W. V., 1951: On Carnap's Views on Ontology. In: QUINE, 1966a, pp. 126—134.

— 1953: From a Logical Point of View. New York: Harper and Row.

— 1958: Speaking of Objects. In: QUINE, 1969a, pp. 1—25.

— 1960: Word and Object. Cambridge, Mass.: M. I. T. Press.

— 1966a: The Ways of Paradox. New York: Random House.

— 1966b: Ontological Reduction. In: QUINE, 1966a, pp. 199—207.

— 1969a: Ontological Relativity and Other Essays. New York: Columbia University Press.

— 1969b: Ontological Relativity. In: QUINE, 1969a, pp. 26—68.

— 1969c: Existence and Quantification. In: QUINE, 1969a, pp. 91—113.

RESCHER, N., 1969: The Concept of Nonexistent Possibles. In: RESCHER, N., Essays in Philosophical Analysis. Pittsburgh: University of Pittsburgh Press.

ROBINSON, A., 1966a: Non-standard Analysis. Amsterdam: North-Holland.

— 1966b: Formalism 64. In: Y. BAR-HILLEL, ed., Logic, Methodology and Philosophy of Science, pp. 228—246. Amsterdam: North-Holland.

RUSSELL, B., 1919: Introduction to Mathematical Philosophy. London: Allen and Unwin.

— and WHITEHEAD, A. N., 1927: Principia Mathematica to *56. Cambridge: Cambridge University Press, 1967.

SCHEFFLER, I., 1967: Science and Subjectivity. Indianapolis, Ind.: Bobbs-Merrill.

SCOTT, D., ed., 1971: Axiomatic Set Theory. Providence, R. I.: American Mathematical Society.

SUSZKO, R., 1967: An Essay in the Formal Theory of Extension and Intension. Studia Logika 20, 7—35.

SVENONIUS, L., 1972: Translation and Reduction. To appear in J. Philosophical Logic.

TAKEUTI, G., 1969: The Universe of Set Theory. In: Foundations of Mathematics, Symposium Papers Commemorating the Sixtieth Birthday of Kurt Gödel, pp. 74—128. New York: Springer-Verlag.

TARSKI, A., 1956: Logic, Semantics, Metamathematics. London: Oxford University Press.

—, MOSTOWSKI, A., and ROBINSON, R. M., 1953: Undecidable Theories. Amsterdam: North-Holland.

THARP, L., 1971: Ontological Reduction. J. Philosophy 68, 151—164.

THOM, R., 1970: Les mathématiques "modernes": une erreur pédagogique et philosophique? L'Age de la science 3, 225—242.

THOMASON, J. F., 1971: Ontological Relativity and the Inscrutability of Reference. Philosophical Studies 22, 50—56.

ULLIAN, J. S., 1969: Is Any Set Theory True? Philosophy of Science 36, 271—279.

WANG, H., 1963: A Survey of Mathematical Logic. Amsterdam: North-Holland.

— 1966: Process and Existence in Mathematics. In: Y. BAR-HILLEL et al., eds., Essays in the Foundations of Mathematics, pp. 328—351. Jerusalem: Hebrew University Press.

WHITE, A. R., 1967: Coherence Theory of Truth. In: EDWARDS, 1967.

Index of Names

Subject Index

Partial List of Symbols